Synthesis Lectures on Engineers, Technology, & Society

Series Editor

Caroline Baillie, School of Engineering, University of San Diego, San Diego, CA, USA

The mission of this Lecture series is to foster an understanding for engineers and scientists on the inclusive nature of their profession. The creation and proliferation of technologies needs to be inclusive as it has effects on all of humankind, regardless of national boundaries, socio-economic status, gender, race and ethnicity, or creed. The Lectures will combine expertise in sociology, political economics, philosophy of science, history, engineering, engineering education, participatory research, development studies, sustainability, psychotherapy, policy studies, and epistemology. The Lectures will be relevant to all engineers practicing in all parts of the world.

Ann-Perry Witmer · Jess Mingee ·
Bernhard D. Scully
Editors

Consilience

Learning About Ourselves by Applying Indigenous Traditions to Western Music and Technology

Springer

Editors
Ann-Perry Witmer
University of Illinois Urbana-Champaign
Urbana, IL, USA

Jess Mingee
University of Illinois Urbana-Champaign
Urbana, IL, USA

Bernhard D. Scully
University of Illinois Urbana-Champaign
Urbana, IL, USA

ISSN 1933-3633　　　　　　　ISSN 1933-3641　(electronic)
Synthesis Lectures on Engineers, Technology, & Society
ISBN 978-3-031-58398-8　　　　ISBN 978-3-031-58399-5　(eBook)
https://doi.org/10.1007/978-3-031-58399-5

This Springer imprint is published by the registered company Springer Nature Switzerland AG
The registered company address is: Gewerbestrasse 11, 6330 Cham, Switzerland

Paper in this product is recyclable.

Foreword

When Ann Witmer first described the Consilience Project to me, I was very intrigued. The project, bringing together music and technology, resonates strongly with my own upbringing, raised by a professional musician and music professor mother, and a professional engineer father who himself was an amateur musician. As I finished high school, I considered the fork in the road between music and technology, choosing the latter with the study of engineering, although music has been an enduring and significant component in my life. I have been a Professor of Chemical Engineering at Queen's University for 34 years, with a cross-appointment to the Dan School of Drama and Music in recognition of my work in innovation, entrepreneurship, and creativity. I am a semi-professional musician, playing flute regularly with the Kingston Symphony Orchestra, as well as jazz in a local quartet, and I also play in an amateur adult flute choir that spans a broad range of both ages and proficiencies. This range of musical pursuits mirrors the range of discussion provided by Bernhard, Jess, Christian, Ariel, Kene, and Alexandra, and is linked to the philosophical questions: What is music for? In what contexts is participation allowed or encouraged? Who decides? How do regional musical art forms evolve? Who contributes? How does this evolution take place across vast geographical ranges, as is the case with jazz and classical music?

These questions, which we might realize more readily with music, come to mind equally with the practice of engineering, and the design and deployment of technology. To what extent does this depend on the context of problems to be solved, and designs to be implemented?

These questions lie at the heart of Contextual Engineering and underscore the importance of The Consilience Project and this book.

Being a musician provides a helpful perspective and appreciation of creativity and expression, within a wide range of musical forms and conventions. Play pick-up jazz at the annual Canadian Chemical Engineering Conference, where participants come from across Canada, and you immediately see that there is a shared understanding of form across vast geographies, from Circle of Fifths progression and 8-bar AABA jazz form or 12-bar blues, to the unspoken etiquette of improvising—"get in, do your improvisation, get out and let someone else improvise." One might ask whether there is, or should be, a similar mechanism for this participatory co-creation within engineering. And thus, the process of consilience continues. How are pressing problems or needs in a community identified? Do the engineers involved, often coming from outside the community, take the time to understand the context? Is the identification of impactful problems to solve participatory and engaging? Is the place-based knowledge used to guide the problem and the evolution of a solution? Are placed-based knowledge-holders engaged in a meaningful way?

I have always been intrigued with the parallels between engineering and music as creative pursuits within pre-defined frames and associated cultural norms. As such, I was quite struck by the reflections of the classical musicians contributing to this book, and the rather stark perception of "social distance" between the classical music form, with all its richness, and the extent to which it's fully linked to the communities in which it is performed. This is an existential question, particularly given the statistics shared by Bernhard about the proportion of US adults listening to classical music. This concern is front of mind for many, if not all, symphony orchestras in Canada and throughout North America, and it is spurring innovative programming and performance mediums, often involving dispensing with some of the associated rigid etiquette.

The Dunin-Deshpande Queen's Innovation Center, which I helped found more than ten years ago, champions and works to build "changemaker" capacity—in social innovation and entrepreneurship—seeking to catalyze the rate at which people unlock and realize their potential to make a positive impact in the world. Our work is focused both in our region and globally. Much of the work is grounded in thinking in systems—Systems Thinking—with components and interconnections that bring together people and technology, with synergy or emergent behavior from the interaction of people and things. Human-centered approaches—contextual approaches—popularly characterized in the front phases of the Design Thinking framework—have been a core part of our approach as well, emphasizing identifying impactful problems, and the importance of engagement and understanding why people do what they do, before launching into further problem definition and solution. Design Thinking as Systems View—the recognition that ultimately, it is the interaction of people with agency, and the physical world around them, in which solutions are realized, hopefully with positive societal benefit, while mitigating negative impacts. When presented with a new management technique, one is often skeptical, asking, "Is this the latest management fad?" And certainly, there is a significant amount of hype around the Design Thinking approach. However, regardless of what one calls the approach, the essential and enduring element is the front end "contextual" approach—ethnography, observation, engagement, and insight, leading to empathy and shared understanding.

At the same time, there is growing interest in Systems Leadership, or human-centered leadership, bringing together the contextual approach with explicit strategies for building community engagement, and disciplined implementation. This is an emerging field [Kennedy School, 2018], bringing together existing methodologies.

Central to these approaches is the importance of context, emphasizing the importance of the consilience in this book.

In his reflective essay from 2010, Leo Marx reminds us of the etymology of the word "technology" from the 17th century, combining the Greek root "techne" (art or craft) with the suffix "ology" (branch of learning) (Marx, 2010). He notes that at that time, "technology" was a branch of learning concerned with the "mechanic arts." In the 1930s, "technology" became used widely to refer to an object rather than a knowledge. The emergence and evolution of computer-based technologies have further strengthened the popular use of the term as a reference to an object, often abbreviated as "tech."

The study of the mechanical arts could be more readily imagined in regional contextual ways, linking strongly to place-based knowledge and know-how. But surely this provides the context both for engineering solutions and for musical performance, the latter being shaped significantly by the nature and evolution of instruments influenced strongly by place.

The success of an engineered solution will depend on the emergent behavior that arises from the interaction of people and technology. But in popular idioms, "technology" is increasingly seen as an entity in and of itself, i.e., an object. As Marx (2010) has noted, we must be careful not to ascribe to "technology"—as an object—agency, which is ultimately a human quality. One of the lessons to be learned from Contextual Engineering is the importance of context approached with humility, acknowledging the need to work with and advance the important role of agency in the community, in individuals, and collectively, representing the context into which engineering solutions are introduced.

Part of the value of the journeys shared in this book lies in reminding ourselves to check and reflect on our own contexts. This stands out in the reflections of Bernhard, Christian, and Jess about the perceived structural constraints inherent in classical music. One might ask, who are the guardians of the genre? In what ways does classical music become part of communities, and participatory? Why has such a highly structured art form emerged?

This book provides examples having sharp contrasts between music form and community participation—amongst the Aymara people, the Mende, jazz, and the classical performances, both observed by the contributors, and in their personal reflections.

The importance of Consilience: Learning About Ourselves by Applying Indigenous Traditions to Western Music and Technology lies both in establishing in permanent form the consilience between music and engineering, and in the sharing from a diverse group of musicians and engineers of reflections on the road to establishing this consilience.

Ultimately, what leaps to mind from the journeys shared in this book is the inherent creativity in communities, whether in music or technical solutions, understood and nurtured with humility and acceptance, as an invariant and essential force, and an integral part of the human spirit. The perception of and empathy for context are surely elements that will continue to distinguish a human-centered understanding and appreciation of the world around us, in a sea of artificial intelligence and machine learning.

In closing, this book is a collection of journeys that I very much believe you will find engaging and thought-provoking, challenging your own preconceptions, suggesting new ways in which to make a positive impact in your personal and professional lives, through music and technology, and in other pursuits.

A pair of wings, a different respiratory system,

which enabled us to travel through space,

would in no way help us,

for if we visited Mars or Venus

while keeping the same senses,

they would clothe everything we could see in the same aspect

as the things of the Earth.

The only true voyage, the only bath in the Fountain of Youth,

would be not to visit strange lands but to possess other eyes,

to see the universe through the eyes of another, of a hundred others,

to see the hundred universes that each of them sees, that each of them is;

and this we do, with great artists;

with artists like these we do really fly from star to star."

Marcel Proust

La Prisonnière

P. James McLellan
Queen's University
Kingston, ON, Canada

References

Dreier, L., D. Nabarro and J. Nelson, Systems Leadership for Sustainable Development: Strategies for Achieving Systemic Change, Harvard Kennedy School (2019).

Marx, Leo, Tehchnology The Emergence of a Hazardous Concept, Technology and Culture 51(3), 561–577, July 2010.

Acknowledgments

The Consilience Project would not have been possible without our collaborators in Bolivia and Sierra Leone, who kindly hosted us throughout our journeys and in many cases, contributed to this book. Their wisdom and perspectives have no doubt enriched our understanding of the consilience of music and engineering, and it is difficult to express our gratitude for the generosity and kindness they showed us. We express special gratitude towards the communities we visited in both countries, who welcomed us and generously shared their time, space, and traditions with us.

J. M. adds: I would not be where I am today without the exceptional guidance of Ann and Bernhard, who have been incredibly influential to my personal and academic growth, be it finding my voice in music, or honing my Contextual Engineering skills. Furthermore, I am immensely grateful to both the UI Horn Studio for receiving me with open arms and supporting my pursuit of horn as a non-music major, and the Contextual Engineering Research Group for shaping my critical thinking mindset through years of thought-provoking discussions. I would also like to acknowledge and thank Jason Finkelman and the Improvisers Exchange Ensemble for forever changing the way that I approach music and continuing to broaden my collective improvisation skills and my exposure to new kinds of music. And finally, I would like to thank my father, Robert Mingee, and my horn colleague, Joseph Goldstein, for their thoughtful assistance in polishing my chapter towards a coherent and meaningful message.

B.S. adds: I would like to thank my family: Sarah, Ellie, Abby, and Maddie. Without their love and support none of this would have been possible for me. I thank Ann for her friendship and for encouraging me to take the leap into The Consilience Project. I have learned and continue to learn immensely from her. I thank Jess for having the wherewithal to introduce Ann and me, for our continual friendship, and for continuing to inspire me in so many ways. I want to thank the UI School of Music for supporting me through the entire project. Professors Tito Carrillo, Jason Finkelman, Lawrence Gray, and Director Jeffrey Sposato especially helped me with various aspects of the project. I want to thank Mateo Hurtado de Mendoza Sanchez and Leon Lewis Nicole for their help and guidance. I also am hugely grateful to all those who were kind enough to read through my chapter draft and provide such meaningful suggestions: Jeffrey Agrell, Lin Baird, Eli Fieldsteel,

Daniel Friberg, Douglas Hill, Joanne Minnetti, and Darcia Narvaez. I want to thank all of my students at the University of Illinois for their continual support, understanding, and encouragement.

A-P. W. adds: I have been incredibly fortunate to work with a thoughtful, compassionate, and moral group of students at the University of Illinois who collectively participate in the Contextual Engineering Research Group. This group inspires me each day to continue to challenge the status quo in pursuit of a just and equitable approach to engineering through a better understanding of those for whom we design. In a world that is threatened by civil conflict, environmental crises, and an ongoing adherence to that which is comfortable and familiar in our approach to the world, it has been refreshing and inspiring to step outside our comfort zones and ask ourselves what we should—and shouldn't—do to make the world a better place. Jess is but one example—albeit an exceptional one—of the deeply thoughtful scholars in CERG and I owe them a lifetime of thanks for making The Consilience Project possible through their critical thinking and wisdom, as well as for connecting me to Bernhard. Together, the three of us gained an understanding of the value of place-based knowledge and traditions that we can only begin to convey through this book.

Contents

Introduction—The Consilience Project

1

Ann-Perry Witmer, Jess Mingee, and Bernhard D. Scully

It's curious that our team was more than halfway through our research endeavor before we encountered the word *consilience,* a term unfamiliar to any of us but which thoroughly captured the work we were undertaking. Consilience is a conjoining of principles from different academic disciplines to form a more comprehensive theory for broader application. Originally coined by 19th century scientist William Whewell to "describe the informal process by which scientists become convinced about the reality of a new scientific principle or phenomenon," it was extended in the late 1990's by E.O. Wilson to refer more generally to "the convergence (of understanding) between different areas of knowledge" (Segerstrale, 2015).

This convergence already had presented itself to us, as the three of us began comparing notes about our own experiences with music and technology, only to find that all of us were feeling as though critical considerations of elitism, access to knowledge, and bias creep into the decision making that traditional practitioners of our crafts take for granted in advancing our scholarship and art.

That's what led us to collaborate in the first place. As members of academic disciplines endemic to the privileged world—namely classical music and cutting-edge engineering

A.-P. Witmer (✉)
University of Illinois Urbana-Champaign, Urbana, IL, USA
e-mail: awitmer@illinois.edu

J. Mingee
University of Illinois Urbana-Champaign, Urbana, IL, USA

B. D. Scully
University of Illinois Urbana-Champaign, Urbana, IL, USA

© The Author(s), under exclusive license to Springer Nature Switzerland AG 2024
A.-P. Witmer et al. (eds.), *Consilience*, Synthesis Lectures on Engineers,
Technology, & Society, https://doi.org/10.1007/978-3-031-58399-5_1

design—we recognized that our elite knowledge may be limiting our understanding of the wealth of intellectual and artistic resources that lie hidden from the industrialized world in rural, Indigenous, and alternatively developed populations. We asked ourselves whether our perceptions of superiority in the arts and sciences may be altered through exposure to people whose place-based knowledge, evolved over time and without extensive external influence, may supplement our own assumptions of functionality, beauty, sustainability, and creativity.

This was the quest we pursued during our year-long collaboration of disciplines that—while recognized as sharing a relationship of thought and attributes (e.g., Charyton & Snelbecker, 2007; Johannsen, 2004)—rarely has been explored from an application perspective. The questions that drove us were these:

1. Why are certain instruments, exclusive to Western classical music, considered elite and scholarly when barely 1% of the US population listens to any compositions of which they may be a part (Nielsen Music, 2020)?
2. Why do Western engineers apply technologies appropriate to their own society when addressing the needs of people in non-industrialized settings?
3. Do we short our own disciplines by blinding ourselves to the diversity, creativity, vision, and skill that lie outside our own experience?
4. How might music scholars and technologists modify their understanding of expressiveness and functionality by incorporating Indigenous or place-based experience into their work?

To answer these questions, our group decided to spend some time together, learning about each other and how we approach our crafts, while purposefully challenging each other regarding the practices and standards that we've been trained to uphold. Then, we traveled together, choosing the diverse locations of the Altiplano in Bolivia—largely inhabited by people of Aymara descent—and Sierra Leone, which has become a cultural mosaic of African traditions cultivated through its role as a repatriation site for freed slaves from the Caribbean, Britain, and the Americas. Amid the swirl of African identities, we particularly focused on the Mende and Temne ethnic traditions of Western Africa. These two locations prevented us from locking into a particular context as representative of the differences between non-industrialized knowledge and our own by providing a contrast in approach to music and technology that is unique to people and place.

The results? This book captures the insights, experiences, and consilience of understanding generated by this unusual inquiry, not only from our own perspectives but from the perspectives of the Bolivian and Sierra Leonean artists and engineers with whom we interacted. Our goal was not to educate others, but to *learn* from others about ourselves and how we limit our own practices. It is our hope that our inquiry will help others, as well, to expand their own definitions of music and technology by incorporating new dimensions of understanding to eliminate the impediments of privilege and elitism, which too often inform the practice of our crafts.

This text compiles the thoughts and experiences of the three principal investigators of the Consilience Project, along with colleagues from the US, Bolivia, and Sierra Leone who joined us on our travel. The reflections in these essays begin to capture the insights that emerged as we explored place-based traditions, reflected individually and together on what we saw, and situated our understanding in the current scholarship. It is impossible to capture the richness and complexity of all the conversations that unfolded during travel among the three principal investigators, ranging from scaffolding of insights across disciplines to energetically challenging each other's interpretations of what we saw.

A note about our terminology: references to Western or Industrialized societies are intended to represent the culture and values of what the International Monetary Fund calls developed or high-income countries. We reject the terminology of "developing countries" as judgmental and normative, choosing instead to refer to these societies as non-industrialized to distinguish them as outside the globalized economy of production and consumption.

References

Christine Charyton & Glenn E. Snelbecker (2007) General, Artistic and Scientific Creativity Attributes of Engineering and Music Students, Creativity Research Journal, 19:2-3, 213–225, DOI: https://doi.org/10.1080/10400410701397271.

Johannsen, G., 2004. Special Issue on Engineering and Music—Supervisory Control and Auditory Communication. *Proceedings of the IEEE*, 92(4), pp. 583–588.

Nielsen Music/MRC Data's U.S. Midyear Report reveals that amid unprecedented challenges, total audio consumption grew 9.4% over prior year. (2020) Billboard. Available at: https://www.billboard.com/2020-nielsen-music-mrc-data-midyear-report-us/ (Accessed: 02 January 2024).

Segerstrale, U. (2015). Consilience. In: Bainbridge, W., Roco, M. (eds) Handbook of Science and Technology Convergence. Springer, Cham. https://doi.org/10.1007/978-3-319-04033-2_3-1.

Part I
Perspectives from Music

Re-visioning the Meaning of Music

2

Bernhard Scully

As someone from the admittedly dominant position of Western society, I have always felt very comfortable in the classical music world. Before becoming acutely aware of my privilege, I didn't understand why so many people didn't listen to classical music, or even more narrowly, music for horn, or why it is practiced by such a remarkably low number of people almost entirely from the same culturally homogenous sector of society. It is now much clearer to me why this is the case.

The horn, my instrument, is deeply embedded in the classical music tradition and has rarely ventured outside of its aesthetic and value system. I've been blessed to perform horn at the very highest levels, in professional symphony orchestras, with the Canadian Brass Quintet, and as featured soloist on the classical world's most vaunted stages. But the performances and music making in which I participated during The Consilience Project marked the first time in my life during which I made music with people who were not Western musicians. And this new exposure to music-making was among the most transformational musical experiences of my life. Our travels allowed the horn—an instrument that ultimately is genre-neutral despite its identity with Western classical music—to be brought to the Bolivian Altiplano and Sierra Leonean wooded countryside where it was given a chance to expand its horizons and redefine itself. We hope to show that by embracing place-based and person-based values, it is possible to branch away from deeply entrenched status quos, learn things previously unimaginable, and create new pathways in these fields.

B. Scully (✉)
University of Illinois Urbana-Champaign, Champaign, IL, USA
e-mail: bscully@illinois.edu

A.-P. Witmer et al. (eds.), *Consilience*, Synthesis Lectures on Engineers,
Technology, & Society, https://doi.org/10.1007/978-3-031-58399-5_2

This project did not just happen out of thin air. In the Fall of 2018, I decided to incorporate improvisation—the spontaneous creation of music in real time—and composition more in my pedagogy and research. I wanted to work toward a more holistic identity as a musician while challenging many of my implicit values, and I wanted my students to expand their skills and go in directions toward more complete music-making as well. So, when school began that fall, I decided to start consistently implementing improvisation in my lessons. This formed a pathway to teach music in a less hierarchical, more conversational way, one that was centered around each individual student's musical ideas and feelings. It was also incredibly fun. This student-centered teaching approach, which puts improvisation and composition at the forefront, contrasts sharply with the more common standardized, traditional approach to teaching which focuses on instrumental technical proficiency and the rendering of already-written musical compositions.

Jess Mingee was an incoming freshman engineering student at UI that year who demonstrated a strong interest in horn performance, so I accepted them into my horn studio. In their first lesson, I asked them to improvise some music with the notes of a scale they had just played. At first, they resisted because it conflicted with their considerable experience in traditional music lessons. Our push–pull continued over the next few lessons, but by mid-semester Jess was comfortably improvising and even beginning to play their own compositions, inspired by their improvising. Even though I had pushed them way out of their comfort zone, they started to find value in their own creative musical voice. What was happening was more than a musical change. It was a fundamental change of consciousness associated with making music.

The following year, Jess and I began having conversations around concepts like decentering curricula and creativity. They mentioned their mentor, Ann-Perry Witmer, a Contextual Engineer with a background in civil engineering, and how they felt the two of us really needed to meet because we were having the same conversations but in different disciplinary contexts. From that introduction blossomed The Consilience Project. Equally important, we learned so much from each other and our many collaborators as we interacted with societies ignored by much of the industrialized world.

Venturing to Bolivia

In Bolivia we partnered with Christian Asturizaga, a Bolivian violinist who is renowned for his creative improvising in the genres of jazz and traditional Bolivian music, as well as serving as concertmaster of the Bolivian National Symphony.

We also joined forces with *Universidad Privada Boliviana* in La Paz and *Fundación Ingenieros En Acción* (known in the US as Engineers in Action), a Bolivian non-government organization, who were working with the University of Illinois on a technology investigation of the Indigenous Altiplano region of the country. Field Engineer Xiomara Echeverria and FIEA office assistant Andrei Xavier De la Barra accompanied us, conducting their own surveys in partnership with Ann while assisting us with language and culture of the Indigenous Aymara people, an ancient society of Andean origin that thrived contemporaneously with the Inca.

Our trip began in the capital city of La Paz, located high in the Andes Mountains at three thousand and five hundred meters above sea level. We traveled by car to the Altiplano, the most extensive high plateau on the planet outside of Tibet and much higher than La Paz in altitude. The area is windswept with large open spaces, few trees, and the Andes mountains surrounding us in the distance. This is the Bolivian motherland of the Aymara people, who have called the area home for nearly one thousand years. The modern Aymara with whom we interacted have retained much of their ancestral culture, and while many of the men speak the Bolivian language of Spanish, women and children still use the Aymara dialect, creating a communication barrier even for the Spanish speakers among us.

Despite having vast knowledge of traditional Bolivian music, Christian confessed that he had never been to the Aymara communities in the Altiplano and had little to no knowledge of their Indigenous music and culture. This surprised me, until I realized that I had never had any meaningful contact with Indigenous people in the US nor had any knowledge of their music and culture. We were essentially on equal footing when it came to experience with Indigenous people, however I was less prepared than Christian to absorb this new music because my musical travels had never brought me to the nation of Bolivia, let alone its Indigenous quarter. Everything was entirely new to me, from the building construction to the limited resources to the very instruments themselves.

Over the course of six days, our team visited and stayed in several villages, traveling together as musicians and engineers to explore our shared and individual interests. In each village, Christian and I would introduce ourselves by drawing out our instruments and freely improvising together, while the technology team roamed from house to house, asking questions and looking at practices and devices the residents used in their everyday lives.

In the village of Checa Belén, we were met by the elders in the town center when we arrived. They asked about our objectives and seemed delighted to learn that some of us wished to hear their music. The village had many musicians who were eager to share their music with us, and they quickly began to assemble in the center, bringing drums and flutes of various types to play for us. After an introduction to their music, the Aymara musicians asked us to play some music of our own, so Christian and I got out our instruments and began playing various styles and pieces of Western music.

We started with some North American jazz, playing tunes like "Summertime" and "Autumn Leaves." Our assumption that jazz would be received well proved correct, since we noticed the musicians engaging through nods and foot taps, and even occasionally picking up their instruments to mimic a strain. We then played some classical music, including a few duets by Mozart, and found it generated an entirely different response. This "high art" music was met with polite smiles but no engagement, as people began turning to each other to chat or stood to wander off. Then we performed several traditional Bolivian tunes—including "Cantarina" and "Collita"—that really brought the house down. These are songs that all Bolivians know well, so the Aymara felt comfortable singing and clapping along as we played.

In the late afternoon, a man came to the community center with a long cylindrical wooden instrument called a *moseño*. The *moseño* is a type of large flute, what could be called a cross flute. It has a rich, powerful, resonant sound when placed in the hands of a skilled player. So it was a special treat to hear one being played so expertly, and he showed us a traditional song that deals with lost love. Jess, Christian, and I all attempted to play the melody with him on our respective instruments, but the *moseño's* tuning was completely different than our classical orchestral instruments, and I found the rhythmic stylings near-impossible to imitate. As we sat in that cavernous community room together, I felt as though I had stepped into an entirely new world of sound, style, and rhythm, completely out of the realm of music to which I was accustomed. I had already transcended my own musical knowledge by emerging into a musical landscape unlike any I'd ever encountered. With the *moseño* player's permission, I have since taken the melody that was taught to us and used it in improvisational settings and in lessons.

That evening, all the town's musicians and many villagers convened in the center for a great celebration of music. One ensemble performed together on various sizes of *moseños*, all pitched differently and in different tessituras to allow for intricate harmonies and rhythms to emerge from what we could only describe at the time as a cacophony. A drummer, the rhythmic conductor of the ensemble, played a steady beat and regulated the musicians through eye contact and swings of his mallet. Another group—like the first, composed only of men—played in a *zampoña* ensemble in a similar fashion. The *zampoña* is like the pan flute, with subtle differences. It comes in various sizes all pitched differently, all in different tessituras just like the *moseños*. We were all mesmerized as they played tune after tune for us and the village audience.

To our great delight, one villager possessed and played a Western trumpet. He brought his instrument to the gathering and sat down near us to play for the townspeople. He played several traditional Bolivian songs, some of the same songs Christian and I had played earlier in the day. In all my life I have never heard a trumpet played like this. His tone was very intense and filled with passion and power as he played the tunes, a very different aesthetic from my classical brass training, and I was enamored with it (Fig. 2.1).

I tried to imitate his sound in mimicking his tunes, but I grossly overestimated my abilities and simply couldn't capture either the sound or the rhythm of what he had just performed. Sensing that he was the master and I was the student, the trumpeter played another tune to demonstrate his music. Christian recommended we play along with him to better integrate. When he stopped playing, I continued and attempted to improvise on the tune and add to it. But when I looked up, he was glaring at me. With a contemptuous look and a sigh, he lifted the trumpet to his lips and demonstrated the original playing once more, effectively saying, "No, it goes like this, and I'm not interested in

Fig. 2.1 Checa Belén trumpet player makes music with Bernhard and Christian

your improvising." In retrospect, his reaction makes sense to me, though it puzzled me at the time. Improvisation is not a part of the music they were sharing with us, even if it felt unscripted and free to our ears. The music was rather an evolving tradition, handed down aurally from generation to generation, changing as new members joined and older members left. It was a living ecosystem of their ancestral knowledge, and musical syncretism through improvisation was not welcome. In that moment, I needed to respect that boundary, and I hadn't.

This was my first Consilience Project encounter of an instance where there was a cultural commonality, yet a total divide in the way it was being executed. On the surface, musical collaboration seemed similar to my own experience as a musician, but the underlying values that I violated with my expressive interpretation ran deeper than I could have perceived merely through shared performance. I have heard many great Western trumpeters in my life, including classical, jazz, and commercial artists. I had never heard the instrument played as it was performed that night. I could not imitate it and would need to spend much more time making music with him to even come close. I heard something and learned something new and was inspired. My assumptions about what are the parameters of "good" brass playing were markedly expanded that evening. In other words, that which I considered to be *high-quality standard* music completely violated the trumpet players' standard. Rather than performing as the expert musician introducing improvisation to the trumpeter, I needed to perform as the student learning his art from his definition of high-quality standard.

I observed an interesting parallel to this notion of Western "superiority" when looking at domiciles in one of the communities we visited. Nearly all of them were built of adobe. There were also several modern domiciles that stood in stark contrast. Interestingly, no one was living in these domiciles. We asked the members of the community about this, and they said that the Bolivian government, as a gesture to gain favor with the community, built these structures for the community without consulting the residents. They did not think to ask what kind of technology the community already used and preferred. The buildings were plumbed for water that the community did not have, so sinks and tubs were used for storage instead. The domiciles were also too cold to reside in for most of the year and instead were used only on rare occasions when it was extremely warm. All the modern, complex technology infused in these structures was effectively useless within the residents' context.

A few days after this experience, we traveled to the village of Culli Culli Alto, an Aymara town known for its cultural museum, which featured an ancestral Aymara burial site. This burial site was remarkable: all of the tombs (*Chullpares*) were constructed above ground, made of a special type of adobe that stood up to centuries upon centuries of brutal climatic conditions. The museum building in which we met and slept displayed artifacts from this burial ground, which had been in existence for many centuries before the Spanish colonized the region (Fig. 2.2).

The construction of these burial vaults was a strong reminder of the value that resides within place-based knowledge derived from generations of technical development. The museum groundskeeper said scientists have tried to recreate the specific mix of adobe material from which the tombs were constructed but have had little success, as the process for producing this adobe had been lost once the area was colonized. Surprisingly, and in

Fig. 2.2 Chullpares in Culli Culli Alto

conflict with our Western attitudes about burial sanctity, many of the tombs had been scavenged by modern Aymara for use in constructing cookstoves because the material was so durable.

I asked one Culli Culli Alto musician about the history and tradition of their music and if there were specific tunes that were passed down and replicated through the generations. He said music was handed down through performance, and no attempt was made to preserve a particular sound or song. Instead, he said, modern Aymara musicians play songs similarly to their ancestors but changed them slowly over time, depending upon the talent and interpretation of the musicians who played them. Every new musician contributed new sounds and ideas to what was already being played and the tradition evolved that way, collectively and collaboratively.

Imagine if our Western classical music were regarded the same way: no formal learning or training praxis and no written scores. Instead of conserving music, the Aymara treat music as a living, evolving aural tradition. Younger men in the town who express an interest in music are recruited into the ranks of ensembles and allowed to integrate into what the older players are performing. For the Aymara, music is not an academic preservation of tradition but a living, growing part of life.

The villagers in Culli Culli Alto were clearly incredibly proud of their music, in part because it is a collaborative place-based art. Juxtaposed against classical music, which is created by singular individuals and remains changeless over time, the Aymara's music-making is a method of social bonding that is included at the center of community gatherings. It is, in fact, an ever-present part of the social fabric, integral to the community and all its traditions.

Christian had left our group by the time we visited this town, so Jess and I interacted with their *moseño* ensembles much the way we had in other towns. Instruments weren't always the same—one bass *moseño* was so large that it required a sort of bocal, like one found on a bassoon, to reach the embouchure—but the sounds were becoming familiar to our ears. The musicians asked to hear some of our music after they played for us, and we performed some classical pieces as well as modern movie themes like the John Williams lineup of Star Wars, ET, and Raiders of the Lost Ark. In my mind, this was accessible music that would spark interest among the listeners. But a group of women were sitting along the far wall, wrinkling their noses, and whispered together in the Aymara dialect. Translated later, the words they were saying were, "What good is this music? You can't even dance to it." This indeed put into perspective my Western assumptions about what makes music accessible and appealing... or not.

We invited one *moseño* player to try out the mouthpiece of my horn. As he placed the mouthpiece to his lips, he drew in a deep breath and blew for all he was worth, much the way he would blow into a *moseño* to generate the Aymara sound. After several tries and a bit of coaching he produced a classical mid-range sound, and after a few more tries and demonstrations, he was able to take this classical horn sound to a few other notes. As I put the horn away, I decided to give him the mouthpiece, and if I would have had an extra horn on hand, I would have given it to him too. I have often wondered since that time if he is using the mouthpiece, either possibly after getting his own horn or using it on his *moseño*, or possibly in some other capacity. Alternatively, is he laughing at the mouthpiece, thinking it is a silly device that produces an inadequate sound? I look forward to the day when I can meet him again (Fig. 2.3).

Before leaving Bolivia, Jess and I joined Christian for a few music events in La Paz that aligned much more closely with our performance experiences in the US. I joined Christian and three of his jazz colleagues to play at the premiere jazz club in La Paz, Thelonius. They allowed me to select half the set list, so I went with pieces with which I was familiar. They also included their own tunes, almost all with a specific Bolivian flavor, different from anything I had played before. When we assembled to rehearse before the show, it was obvious that these musicians were world-class artists. The rehearsal went smoothly and we were all able to integrate seamlessly. The most wonderful highlight for me that evening was when Christian and I decided to use the melody we learned from the *moseño* player in Checa Belén in a free improvisation, as a gesture of celebrating our experience with the Aymara people. We freely improvised with each other on the Aymara

Fig. 2.3 Bernhard demonstrates French horn embouchure and air speed to a Culli Culli Alto musician while Xiomara provides language translation

melody and rhythms for the better part of fifteen minutes: no music, no plan, everything in the moment. We completely lost ourselves in the music-making and our performance reached a level that I had never ever before experienced in front of a live audience. The audience rose to a standing ovation as we ended the piece, the most intense response we had to any of the music we played that night.

But did we do the Aymara music justice? I can't help but recall the reaction of the trumpet player to my improvisation, and I must recognize that this free improvisation may have satisfied *our* musical souls even though it likely would conflict with the soul of an Aymara musician. How ironic that the height of my musical experience lay not in adopting the Aymara music we had been studying for the previous week, but in modifying that music to address our own values and sentiments. This begs the question of whether that performance was truly a gesture to share something we had learned from the Aymara people with the audience, or whether their tradition was lost in adapting it to our musical setting.

Stepping out of this question for a moment, I must acknowledge that jazz as an art form is particularly adept at integrating virtually any other kind of music. The reason for this is because of its creative improvisatory base. Improvisation is foundational to jazz. Because improvisation itself is a mechanism towards transdisciplinary collaboration and integration, jazz has this built into its core. It is not surprising to me that my most immersive musical experiences, where I learned the most in both Bolivia and Sierra Leone, often included improvisation and often were directly related to the jazz ethos. The Aymara tradition was not set up to integrate with other music, so I was mistaken to assume the trumpet player in Checa Belén would be receptive to a syncretic ethos.

The next morning, we all convened one last time to perform our traditional classical concert at the Symphony Hall. Christian, his symphony-organized chamber orchestra colleagues, Jess and I performed a concert featuring works of Mozart and Telemann for a sizeable and appreciative audience. Before the performance, Christian had set up a master class for me to work with brass students from a local music conservatory. The experience reminded me that in Bolivia, classical music is neither studied as rigorously nor regarded as highly as in Western-influenced countries. Christian, for example, is the top classical violinist in the country and concertmaster of the symphony, but he seemed to be more, or at least equally, renowned for his creative improvisatory work. His membership in *Orquesta Criolla* (a La Paz ensemble specializing in traditional Bolivian music), with whom Jess and I recorded several tracks for their newest album while there, seems to generate much more interest than the National Symphony.

On to Sierra Leone

Six months after our return to the US we travelled to Sierra Leone, which was my first trip to Africa. We partnered with Njala University professor Philip Foday Yamba Thulla. Philip is a writer, researcher, editor and Director of the Institute of Languages and Cultural Studies at Njala University. Due to the gradual disappearance of the place-based musical traditions in Sierra Leone, his further purpose was to use The Consilience Project as a launchpad for documenting the Indigenous musical traditions of Sierra Leone.

Sierra Leone was unlike any place I'd visited before. We were obvious outsiders in appearance, dress, and demeanor, though the sheer numbers of people in Freetown allowed us to melt into the background. Quickly, though, we travelled to the rural communities of northern Sierra Leone to experience Indigenous music there, and we became very conspicuous. The first several villages we visited welcomed us warmly, assembling their musicians for concerts and escorting us to seats of honor to listen and engage. I say engage, because music in Sierra Leone is not passively consumed.

The first few performances we witnessed included almost no melodic or singing component, so right away we noticed stark differences in relation to the heavily tonal Aymara music we experienced in Bolivia. The music was highly intense, with interjections of whistles and provocative dance movements. For my Western ears, it took me a while to make any sense of what was happening, especially because, in my field of classical music and the melodic nature of my own instrument, the lack of melody and harmony made me feel like something was lacking. Sometimes, because of the intense dancing, I even felt uncomfortable. I had never heard music like this before, at least not in a live setting, and it felt spontaneous and unrehearsed, even though the performances to which we were witness typically were reserved for royalty and dignitaries. The drumming was elaborate and required great skill, and only select members of the community were skilled enough to play the music. Many different types of drums were played all at once. The dancing was powerful, displaying all types of emotions, and at times was quite sensual. There was no opportunity to bring our horns out and join in, nor do I think we would have wanted to try to collaborate. Instead, Jess, Ann and I opened our minds, eyes and ears, sitting like sponges listening.

At another rural village, we were once again met by all the elders of the community. Three women who were dressed in a special garb of striped red and white presided over the gathering, and we quickly learned that they had the power to sentence people for crimes or even use traditional magic to cast spells and curses. These women also conducted the music, which was in the form of a call-and-response. The vocalist, whom we were told is one of the most highly regarded performers in all of Sierra Leone, sang verses through an electronic megaphone, and the audience responded in song. Call and response, although not typically present in classical music, is one of the most common improvisation-style forms of music-making in the world and is used in Western blues, jazz, and gospel, all of which have roots in African music. The women elders danced before us while playing rhythm instruments, and once again, I felt intrigued but uncomfortable with the closeness and personally attentive nature of the performers. All the villagers then began dancing and we were beckoned to dance as well. Though I resisted initially—dancing with a bunch of people I have never met to music I didn't know in a part of the world I have never been?—I eventually acquiesced. After a few minutes of feeling discomfort on a level that I hadn't felt since my first school dance in sixth grade, eventually was able to ease in. Soon, close dancing felt perfectly natural, and we danced away the afternoon. It was very freeing, and I wish this type of experience was more a part of my musical life in the US.

We had the pleasure of interacting with a popular musician in Sierra Leone who merges hip-hop and pop with traditional and Indigenous music traditions. Kene Muadie is a multifaceted artist who is known for playing Mende music on his accordion, often with a hip-hop spin. Meeting Kene signaled a shift in our journey as now we would be able to take out our horns and make music with him. We gained much more perspective on Mende music as a result. A singer-songwriter who draws upon traditional themes and melodies, Kene accompanies his lovely singing voice with his accordion. His songs were often in a triple rhythmic feel (like a waltz or minuet), always in a major key, and so they sounded much more familiar to our Western ears than did the music we heard the previous day in the villages or in Bolivia with the Aymara. Kene sang a few songs and then we began accompanying him on our horns. His music is delicate and subtle, so our instinct was to play a harmonic underpinning to accompany him which we did for a couple of minutes. But he beckoned to us to take a chorus on our horns. Jess and I each improvised a whole chorus with him accompanying us on the accordion, outlining the rhythm and chords. It was challenging and fun (Fig. 2.4).

The crowning event of our African experience was a visit with the Sierra Leone National Dance Troupe. This group is one of the most highly regarded ensembles in West Africa. Six musicians from the group met outdoors under a spreading Banyan tree to share their music with us. They sang, used four drums, and one *balangi* (a xylophone-like musical instrument consisting of a series of wooden tone bars or keys attached to a wooden frame, with small gourd resonators beneath, played with two beaters). The tones of the *balangi* are not equal tempered, as is the case with a piano, but rather are based in what could best be described as a harmonic series scale. Joining the group was an activist artist who goes by the name King Kuntaman. He sang, recited spoken word, and played a singular drum. His songs address revolutionary topics, often being political protest songs. After he sang, we were then asked to take out our horns and make music together.

Jess and I joined the ensemble with our horns. Unlike in Bolivia, the group asked us to start by playing something we knew, and they would then join us. Jess and I played a slow bluesy groove in duple time, thinking that whatever was going to happen, we wanted to keep it simple. To our surprise, the group immediately took what we set up and began to elaborate on it, adding new rhythms and harmonies that complemented what we were playing. The first four minutes we stayed relatively minimal and slowly built on what we

Fig. 2.4 Jess and Bernhard integrate their horns into the hip-hop-influenced Mende music that Kene frequently performs

started with. At the four-minute mark, there was a sudden and radical shift to the music we were making. The group unanimously turned what we had been creating into one of their own songs with a nearly identical beat and feel but with a distinctly different flavor.

Once the shift happened, the group continued to build on this new music and we were inspired to improvise. The tuning of the *balangi* forced us to think outside our normal instrumental tuning. The singer—the only woman in the group—and *balangi* player would riff for a bit and then I would improvise, call-and-response style. It took me at least three choruses to internalize and then play back the equivalent intervals and tuning. We were able to take on the cultural characteristics of the dance group and they were able to absorb our cultural personas in turn. We had taken the Western classical horn into a musical space it likely had never been and integrated it using all our available faculties. It was a truly transformative experience that became a center point of discussion for Jess and me. As in both the Bolivian jazz club and our jam session with Kene, when both parties were willing and able to improvise together, the artists could immediately integrate, and an entirely new music could be formed in the moment.

The Horn and Classical Music

I have so far made much reference to "classical music", especially in contrast to the Indigenous music and other non-classical music we engaged with on our travels. Classical music is the music that I have spent my entire life engaging with and have built my career upon. Until recently I never even thought to examine it from any but an insider's perspective. That all changed after 2020.

For simplicity, and acknowledging that there are some exceptions, the term classical music in this context is used to mean a type of music originating from the geographical location known as Europe, that grew out of both the ecclesiastical and aristocratic sectors of society, spanning from the Middle Ages into the present day. There are many classical music traditions around the world in countries like Iran, India, and China, but I am not referring to those here. The classical music from Europe is generally considered distinct from popular and folk music traditions, and it does not typically represent Indigenous and minority cultures nor even a majority of Europe. There are a lot of qualifiers when it comes to describing it, but people will tend to recognize it even if they can't put it into words.

The classical aesthetic (sometimes referred to as the Euro-logical aesthetic) largely values such things as harmonic and rhythmic complexity, as well as sonic and tonal "purity". Sonic resonance, balance, and unification are sought after. Precision and accuracy are valued in performance. Because of these attributes, it entails a high level of skill and training to perform in any capacity. Distinguishing it from many aural and vernacular traditions around the world, classical music is largely preserved in the form of compositions or written works which utilize its notational system. With rare exceptions, these works are almost entirely created by singular individuals, composers, and not in group settings—we refer to this mass of works as the "classical canon". One can see how this aspect would significantly juxtapose it against the music of the Aymara or Mende, which are entirely from an aural tradition and created collaboratively. Classical music largely exists unto itself without any words or movement, again being starkly different from the music we encountered in both Bolivia and Sierra Leone.

Critiquing the Canon and the Horn's Diversity Dilemma

There is much debate about how, why, and who gets to choose the makeup of the musical canon. For the record I am a big fan of all the canonical repertoire, and I have joyously performed much of it that is written for horn. Yet the very concept of a canon of knowledge has been highly suspect and critiqued in the last few decades. For every work in the canon, there were countless other works written which have been lost to history or left out for all kinds of reasons. Those who were in power and who made these decisions in terms of which music was included were of course of the dominant caste: the canon has therefore traditionally excluded nearly all the world's general population (intentionally and unintentionally), especially women, LBGTQ+, and BIPOC, and anything outside of the dominant caste.

A common retort to this critique is to say, "Other classical traditions around the world have also been exclusive of everyone outside their cultures, so why should our classical music be any different?" It bears mentioning that one needs to take account of the longstanding relationship that the dominant caste, the practitioners of classical music, has had with the rest of the world. Any possible exclusionary practices may lie more in their desire to protect what they have rather than from any position of superiority. Exclusionary practices based in protecting one's own culture from aggressive outside forces are on the opposite end of the spectrum in relation to practices based in cultural supremacy.

Even if it is true that other cultures have highly exclusionary practices it doesn't change my resolve. It is rather absurd to say that "because other people do questionable things, therefore it is ok for everyone to do those questionable things." If there are oppressive forces such as racism, sexism, classism, ableism, etc. embedded into an artform, then I believe it is necessary to do our best to change this. If other art forms have these oppressive forces in them, I sincerely hope they would want to change too.

One of the most obvious skills that would allow classical music to coexist with traditions from around the globe in a less exclusionary way is through improvisation, but this has been systematically taken out of the classical praxis in favor of gaining technical proficiency to preserve the music of the past. The horn, in fact, is one of the world's oldest acoustic instruments and held its place in a breadth of cultural traditions before it became incorporated into the Western classical tradition. If the goal of the artform is only to preserve the horn's classical history and tradition, then there really is little point in talking about making the artform more inclusive, diverse, equitable, and accessible or even including improvisation in the first place.

Through observation and my lived experience, I now know that it is possible to change. It is possible to transform an art form like classical music to be accessible to all, not just for a very few, and to be representative of all people. I have seen this in my own studio of horns in recent years since integrating improvisation into our routines. And the only way that Jess and I could have had the experiences we did in Bolivia and Sierra Leone was through improvisation. If the goal is to diversify the artform and integrate it into many other types of music, then reintroducing improvisation would be a logical first step. The Consilience Project itself is largely a result of such an effort.

The Syncretism of Bolivian Music

Christian Asturizaga

I have the privilege of serving as concertmaster of the *Orquesta Sinfonica Nacional*, a position earned through training in Western classical music. And yet, I do not consider myself solely a classical musician, partly because my heart also lies with jazz and modern music, as well as with Indigenous Bolivian folk traditions. This is not a contradiction of terms, but rather a representation of the Bolivian understanding of music, an understanding that, while traveling with Bernhard Scully, Jess Mingee, and Ann Witmer, I learned is quite different from the Western classical tradition of musical performance.

Let me explain. I play the violin, and I studied the violin with classical instructors in the European tradition. Through that experience, I learned technical proficiency, as well as the importance of understanding and reflecting the composer's intent through the mastery of technical and expressive skills. But I also was born and raised a Bolivian, where I was exposed to a very different understanding of music, one that encouraged me to express my own intent and expression, one that regards music not as an art to be passively observed but as a participatory art that encourages engagement of both the performer and the listener.

The differences between our experiences were quite apparent the first day Bernhard and I withdrew our instruments from their cases and began to play for the residents of Santari. As residents gathered to listen to us improvise music together, they watched us suspiciously for only a brief moment before joining hands, pulling in Ann and Jess, and circling around in traditional Bolivian dance. Ann noted afterward that I seemed totally at home playing for these people. We sat on a porch overlooking the village,

C. Asturizaga (✉)
Bolivian National Symphony, La Paz, Bolivia
e-mail: christianasturizagah@gmail.com

while people from the community filtered in to stand or sit in the grass before us, sharing their coca leaves with Ann and pulling their guests up to shuffle on the makeshift dance area. It was organic and natural. Though it was the first time Bernhard and I had improvised together, we got along very well and our musical dialogue was really inspired. Still, Bernhard seemed to wonder if the sounds he produced were appropriate for the activity before him, though he gleefully joined in the dancing after setting aside his horn. I sensed in him a clear division between performer and consumer, engaging more comfortably with the community as a co-listener than he did as a horn player in this setting.

In full disclosure, I have not spent a great deal of time on the Altiplano with the Aymara people, but I'm very familiar with their folk tunes and very comfortable playing those tunes on my violin for others. The music I made in Santari may not have employed the same virtuoso technique I would use in the symphony hall, but it certainly matched the expectations of those who listened and danced and clapped along. Bernhard also was not familiar with the Altiplano or the Aymara people, but he appeared uncomfortable with the informality of our performance and the unrehearsed music we made. Did he worry that he was doing it wrong? Did he regard the music as too simplistic, not enough of a technical challenge? Was he concerned that others would judge him based on the fact that his instrument didn't sound familiar to the Aymara ears that heard him? Whatever the reason, he quickly fell back on playing some Classical standards when he picked up his instrument again, and the engagement of the audience rapidly drifted.

By contrast, after we had spent several days together and Bernhard had gained familiarity with the range and timbre of Bolivian instruments, we sat together on a bench near the town square in the municipality of Sica Sica, improvising music together again, this time with a touch of Bolivian air. Passersby stopped to listen, and a small crowd gathered as we played, once more nodding and clapping along to our music. Something had changed since our first performance—Bernhard seemed far more at ease. I can't help but think it was less a change in the music than a change in thinking. Bernhard had begun to adapt his horn to the Bolivian way of thinking about music, rather than trying to fit Bolivian music into his own way of thinking (Fig. 3.1).

This is an important point. I believe there is a syncretism associated with the Bolivian musical experience, whereas the typical Western classical musician is trained to conserve and promulgate a particular sound, a particular form, a particular way of listening to and performing and engaging with music. When that form, that sound, that engagement differs from expectations, the Western musician questions whether it is, in fact, interesting music at all.

I once played and recorded the compositions of a blind musician who dreamed of being elected to the Paris Conservatory, and he composed music but couldn't write it and would have people write it down for him. We looked at his pieces with our string quartet and found them to be very, very repetitive, which is common to Bolivian traditional music. I asked some colleagues their thoughts, and they said we must cut, cut, cut, in a

Fig. 3.1 Christian and Bernhard improvise in the shade of a portico on the village square in the municipality of Sica Sica

classical thinking way, put it in a format that was easier to listen to and work with given the time constraints of performance in a conventionally classical setting. In our effort to conform with Western tradition, we were taking off a lot of what was there, a lot of what made the music Bolivian. It was too different. There is a conformity associated with classical music that prevents musicians from hearing the musicality and understanding the thinking that accompanies other forms of music, such as the repetitive folk songs of the Altiplano. And that conformity makes it difficult for a classically trained musician to understand the benefit of alternative ways of thinking about music. We solved our need for performing and recording these Bolivian pieces by changing our thought about how to perform this music; we created a technique to develop the concept of timbrical improvisation for strings to keep the music interesting. I discovered this need of a different mindset while playing *moseño* and trying to blend my individual blowing and fingerings to match my understanding of the piece with the *tropas de moseños* (community flute ensembles), recognizing that skills are needed beyond codifying the notes of Western scales. These skills include understanding that sounds are produced to complement sounds of other instruments and are performed using fingerings and methods of blowing into the *moseño* to reflect an interdependent sound. One accomplishes these complementary sounds by imitating the movements of the other musicians, rather than following the skills I developed as a player who performs alone. These same skills could be transferred to a different instrument like my violin or Bernhard's horn.

Bernhard and I performed in a jazz club in La Paz while he was in Bolivia, and the experience was very different from our interactions with Aymara musicians. I'm not sure if this is because jazz is a familiar music form to Western performers. We were doing essentially the same thing as we had done while visiting the Altiplano communities and playing along with the local bands, though we were doing so with the very highest quality of jazz musicians on the continent. But I do think there was a great deal of legitimization through conformity with musical standards that made the jazz club more comfortable to the Western performers. Bernhard and Jess also had the experience of recording with "Música de Maestros," a traditional Bolivian orchestra with whom I've performed. This ensemble is experienced in blending Western and indigenous sounds, and the musicians really appreciated the sound quality and contribution of the horn players. The need for legitimization or conformity doesn't exist in Bolivian music, since musicians here are exposed to not only Western classical and jazz, but also traditional bands, all coming together syncretically to create a Bolivian definition of music. We can play the piano, the violin, the guitar, and Indigenous instruments like the *moseño*, and mix them all together to create a sound that is more alive than any of the individual instruments. But you have to be willing to *think* that each of these instruments can blend to create music. That's Bolivian. That's what I mean by the Bolivian way of thinking about music.

Now, I must confess that even Bolivians worry about how their music will be perceived by others. When Bernhard joined our chamber ensemble for a rehearsal and performance, the string instruments were genuinely afraid to play with him. They knew of his stature in the Western classical music tradition, and they worried he would think poorly of them, which led them to play with anxiety about the quality of their performance. I had told them this was a wonderful opportunity to join with a world-class performer in their own hall, playing their own music, but they were frightened. It actually wasn't all that different from the way Bernhard appeared to be feeling when he first tried to play the horn with a *moseño* band, none of whom were classically trained and none of whom had an understanding or appreciation of the type of music he made.

Perhaps if we think of music as binary, as all art forms are, it's a little easier to explain. Music is either tension or relaxation, right? But what determines relaxation and tension? In Western music, relaxation is harmonizing, gentle, classical. Tension is dissonance. But in some forms of native Bolivian music, sounds that in a Western tradition may be considered dissonance and harshness are the sounds that are more expected, so we can consider them to provide relaxation. Gentleness and lyricism, by contrast, are sleepy and therefore produce tension. So we need to recognize that what may please Western ears, what Western classical audiences may seek to consume through passive observation, can sound disconsonant in the context of the Pre-Columbian musical tradition. This shapes the musical tastes of the people we visited. Through syncretism, though, the music that is heard finds a way to appease the tastes of both Western classical and Indigenous Bolivian music consumers. Indigenous brass and *moseño* players provided us with of another fantastic experience while in a gathering we played tunes in turn. With our instruments, we imitated the melodies they played, and then they imitated the ones we played. All of the tunes were popular and tonal so the musical language was technically built upon the influence of Western tradition, making it easy for us to replicate. By contrast, when the horn players tried to play the *moseños* and *moseñadas,* their advanced musical training gave them no advantage, allowing the Bolivian musicians to feel more confident in expressing their own musical thoughts.

This leads me to my final point. While I was traveling through the Altiplano with the Consilience team, we were investigating technologies used to obtain and preserve safe drinking water for rural Aymara communities. Because many of the community residents and I share a national identity, if not a cultural one, they confided in me several times that they felt compelled to say they'd accept any technology provided to them by the visitors regardless of whether they thought they needed or would use it. This is the equivalent, I believe, of saying that one will listen to a Western classical performance whether one likes it or not, simply because a performer is there to share his or her talents. As Bernhard can tell you, though, the Aymara people were quite comfortable in expressing disinterest in his classical pieces on the horn. Why the difference? Perhaps because there is nothing to be gained from listening to another's music, but a technology that doesn't align with needs can always be sold or repurposed, meaning it has a tangible value.

So my advice to those who seek to work with non-industrial communities such as those we visited is to seek a syncretism of approach, incorporating multiple schools of thought to find a solution rather than imposing the Western standard of technical purity upon the users of that technology. Put more simply: Western visitors should adopt the strategy that Bolivian musical thinking uses and merge their own understanding of technology with that of the technical consumer.

Looking at Consilience Through a Mende Lens

4

Kene Muadie

Identifying with Mende music may mean a deep connection to our cultural heritage, a sense of belonging, and a way to express our identity. It could signify a pride in your Mende roots and a desire to preserve and promote the traditions and values associated with Mende music. I chose Mende music as a medium of self-expression because it resonates with my personal experiences, emotions, and cultural background. Mende music provides a platform for me to convey my thoughts, feelings, and stories in a way that is authentic and meaningful to me.

Compared to other music, Mende music may have distinct rhythms, melodies, and instruments that differentiate it from other genres. It may incorporate specific cultural elements, traditional songs, or dances that are unique to the Mende culture, setting it apart from other forms of music I engage with. I particularly appreciate the vibrant rhythms, rich melodies, and lively performances of Mende music. The cultural significance and historical narratives embedded in the music might also resonate with me. Additionally, the sense of community and connection fostered by Mende music could be aspects that make it especially important.

I am able to express myself through Mende music, which contributes to my personal identity as an artist by allowing me to embrace and celebrate my cultural heritage, adding depth and authenticity to my artistic voice. It can also contribute to my identity as a human being by fostering a sense of connection to my roots, enhancing my understanding of Mende traditions, and promoting cultural exchange and understanding among diverse audiences.

K. Muadie (✉)
Freetown, Sierra Leone
e-mail: askiakargbo11151@gmail.com

Mende music, originating from the Mende people of Sierra Leone, utilizes a variety of traditional musical instruments. Here are some instruments commonly used in Mende music:

- Saka—The saka is a type of thumb piano, also known as a kalimba or mbira. It consists of metal or wooden keys attached to a resonator. The player plucks the keys with their thumbs to create melodic patterns.
- Kondi—The kondi is a large thumb piano with metal or bamboo keys. It is played by plucking the keys with the thumbs while holding the instrument against the chest or on a stand.
- Ngongoma—The ngongoma is a type of musical bow made from a flexible stick and a string. The player uses a small wooden stick or their fingers to strike or pluck the string, creating rhythmic sounds.
- Shegureh—The shegureh is a type of instrument made from wooden or bamboo keys suspended over resonating gourds or wooden frames. It is played by striking the keys with mallets.
- Bata—The bata is a set of double-headed drums made from carved wood or pottery. It consists of three drums of different sizes, each producing distinct tones. The drums are played with the hands and sometimes with sticks.
- Bondo Drum—The bondo drum is a large barrel-shaped drum covered with animal skin. It is traditionally played by women during Bondo society ceremonies and other social events.
- Gara—The gara is a rattle made from a hollow gourd filled with seeds, pebbles, or shells. It is shaken to produce rhythmic percussive sounds.

These are just a few examples of the instruments used in Mende music. It's important to note that Mende music is diverse, and the instruments used can vary depending on the specific context, regional variations, and individual preferences of the musicians.

You'll notice that there is no horn in Mende music. In fact, I had never encountered the French Horn before I met Bernhard Scully, Jess Mingee, and Ann Witmer during their visit to Kenema. Through my observation, the horn produces a very peasant and melodious sound, making the rhythm enjoyable to listen to. The instrument is made of brass, which makes it lightweight and easy to carry. I note that horn music has a wide range of tones and can produce an emotional, expressive sound to entertain people. It is my belief that playing the horn is considerably more difficult than other instruments like the accordion, which I use, and mastering the horn requires specialized skills, knowledge and practice. Observing the response of the crowd when the horn was displayed during a recital at Njala University Bo campus, I sensed that the complexity of the horn was widely appreciated and enjoyed, especially when performed in combination with the accordion.

Part II

Perspectives from Design and Technology

The Way We Approach Technology

Ann-Perry Witmer

I am not a music scholar, and make no claims otherwise. My musical credentials start with playing around on flute, cello, and classical guitar, none of which do I even practice anymore, and end with high school music theory class. But as a Contextual Engineer, I am a practiced student of how people approach knowledge systems and interact with devices that address a need (Witmer, 2020). My systems of study typically are physical infrastructures such as community water systems, designed by a few technical experts to address the needs of substantial user populations.

In a past life, I was also a storyteller of sorts, exploring ideas and questions about the way government worked or environmentalists behaved or courts meted out justice, by asking questions, looking for connections, and sharing my findings on the pages of daily newspapers. These erstwhile skills of a previous career served me well during The Consilience Project, collaborating with engineers and musicians to learn about the relationship between Western society and the vast knowledge that lies in the non-industrialized world, mostly ignored and trivialized by us who pride ourselves on our own erudition.

Contextual Engineering

The foundations of Contextual Engineering lie in my work as a water supply engineer, volunteering with Western service organizations to provide safe water access to non-industrialized communities around the world. The more exposure I had to these so-called

A.-P. Witmer (✉)
University of Illinois Urbana-Champaign, Champaign, IL, USA
e-mail: awitmer@illinois.edu

© The Author(s), under exclusive license to Springer Nature Switzerland AG 2024
A.-P. Witmer et al. (eds.), *Consilience*, Synthesis Lectures on Engineers,
Technology, & Society, https://doi.org/10.1007/978-3-031-58399-5_5

"developing" societies, the more aware I became that the measuring stick by which development is assessed is controlled by the high-income, high-influence nations of the world. Alternative approaches to technical design were viewed by my fellow volunteers as inferior simply because they didn't employ the materials and methods used in the US or Europe, and designs created by Western engineers often ignored such critical considerations as maintainability, resources, financial constraints, and societal values and beliefs. Our industrialized-world context, combined with the hubris of privilege, permitted us to view our own technology as superior in all situations and for all people.

When Bernhard Scully and I met through our mutual student, Jess Mingee, we quickly found ourselves echoing each other's words about privilege and power. But Bernhard was talking about music, while I was talking about engineering. What a strange coincidence, I thought as Bernhard described how the music regarded as most scholarly and elite—Western classical music—is among the least consumed genres of musical performance on the planet. Ditto for the technology we create in the rarified air of academia, I thought. How odd that academia has traditionally disdained Indigenous music genres, treating handmade instruments such as flutes and drums as tawdry and unrefined, much the way engineers consider technical designs that employ materials like bamboo and adobe to be vastly inferior to those made of steel or polymers.

Together, Bernhard, Jess, and I decided to explore a few non-industrialized societies for the purpose of learning what academia has ignored regarding place-based knowledge, both in the realms of music and technology. Along the way, I also watched how the three of us reacted to place-based practices, curious as to whether we displayed deeply embedded implicit biases that might explain why the worth of non-industrialized traditions is so commonly ignored in Western institutions.

Observing the Knowledge Holders

Our experiences on the Altiplano of Bolivia were particularly technology-focused because of our collaboration with *Universidad Privada Boliviana,* a Bolivian university, and *Fundación Ingenieros En Acción,* a non-government organization. The three entities were working on a contextualized introduction of climate-friendly technologies for cooking, pumping water, and generating electricity. The intent of this partner effort was not to impose technologies upon the Aymara, who typically live austerely, so much as to understand the needs of the population and the technologies with which they would most comfortably engage. Such an investigation means that our team dispensed with phrases like "you should…" and "it would be better if you…," instead gathering the knowledge of the residents with questions like "how would you like to…" and "what do you do to…" In other words, we sought to gather place-based knowledge from the holders of that knowledge. At the same time, Bernhard and Jess, accompanied by Bolivian maestro Christian Asturizaga, explored music in much the same fashion. With an objective of resisting imposition of a Western musical idioculture, the trio sought to gather understanding from Aymara musicians.

In the cases of both the technological and the musical inquiries, intentions were better than execution, however the challenges were different with each discipline.

Our technical inquiries often were met with suspicion and reticence by the Aymara, who were uncertain of the benefits and drawbacks that would accompany them sharing their private lifestyles with outsiders. Christian later confided that many of the Aymara talked separately with their fellow Bolivian to share that they felt compelled to tell us whatever they thought we wanted to hear. The powerful impact of perceived Western affluence and influence led members of this marginalized Indigenous population to assume we didn't really care about their practices and beliefs, and they would do well to placate us with easy answers to obtain access to our resources and wealth. There were several people who stood out as engaging with honesty and as equals, most notably an elderly woman in Checa Belén who had overseen the installation of the original water system for the community and was open to discussing the process with a fellow woman technician, at least until we asked if we could take a photograph of her to remember the conversation.

Musical demonstrations took an entirely different tenor, during which the Aymara exhibited great confidence and pride in sharing their performances with the visiting musicians. With little prompting, town bands would assemble their members for an impromptu concert for us, demonstrating a complexity of rhythm, tone, and harmonization handed down through generations of families from father to son (women did not seem to play a role in the performance of traditional music, at least among the Aymara communities we visited). Several of the musicians offered to teach their instruments to our musicians, and they wore bemused smiles when the world-class horn player struggled to draw a sound out of their traditional flutes or became downright frustrated when he was unable to produce a rhythm on his horn that they could elicit from their instruments.

What was the difference between the music and technology inquiries in Bolivia? Perhaps the Aymara have a distinctly personal connection to their music form, which they consider superior to any other form or instrument; their technology, however, is not as inherently connected to their identity, leaving them less protective of it or more willing to disavow it when faced with an opportunity for something different. It would explain why visiting engineers feel comfortable disregarding place-based practices in favor of their own, particularly when non-industrialized societies have been indoctrinated over centuries of colonization and globalized influence to identify Western technology as valuable. When the product of that knowledge is an integral part of their identity, though, as the community bands in each of the Aymara villages demonstrated, there is no question about preservation of tradition over Western "superiority."

We also noted that many people we encountered chose not to engage with discussions of technology at all, deferring decision-making to community leaders or others in their families rather than adopting a direct role in identifying technology use, maintenance, and development. Everyone, though, participated in music, whether they performed or listened. I use the term "listened" loosely, because musical performances in the communities we visited rarely were met with passive observation. When the bands played, people danced. When the music stopped, people implored the performers to begin again. Concerts were exhilarating gatherings that sometimes involved copious drinking by both listeners and performers, sometimes involved sudden shifts from one type of music to another in response to the crowd, sometimes simply stopped without any clear cue that the event was done. The performers were not set apart from the listeners but stood among them, and listeners didn't sit quietly and applaud politely, but instead were a constant swirl of activity, conversation, and interaction. A greater contrast with a symphonic concert in the US or Europe could not be imagined.

But this was one culture in one country on one continent in the Western hemisphere. How would these relationships play out in another place, on another continent, on the opposite side of the world? Our visit to Sierra Leone provided an opportunity for comparison.

Admittedly, my primary objective in Sierra Leone was to observe differences in experience from Bolivia, rather than exploring technology. Previous visits to the country had provided me with sufficient understanding of the technical infrastructure, which was heavily dependent upon availability of the two most tangible resources of time and money. The villages and cities we visited in this West African nation were not as heavily steeped in the ancestral traditions that imbued Aymara society, and as a result people were more open to introducing new technical processes, provided they met a need and didn't demand more than people could provide. At the end of the day, the sophistication of technology in Sierra Leone was not significantly different than that found in the Altiplano, but the reasons for simplicity appeared to be more rooted in access than in cultural value.

But music… the music of the rural villages in northern Sierra Leone was as rich and ancestral and participatory as any we encountered on our earlier travels, though the sounds were unique to the society and the performances on another plane. After one of our early observations of Mende music, Jess turned to me and said, "I think I just had my first lap dance." This was because the music wasn't simply music. It was music and dance and song and very personal engagement, all in one. We had the benefit of traveling with Dr. Philip Foday Yamba Thulla from Njala University, who warned us that this music typically is accompanied by a dance troupe and the performers all engage in quasi-erotic movements that are not perceived by the Mende as sexual in nature but can be off-putting to those unfamiliar with the art form. He was correct, as chairs were brought out for Jess and me, and dancers teased us with a shake of a grass skirt in our face, an eye-to-eye invitation to hop up and gambol together, a wave to come join a hip pivot or shoulder dip. At one community, a group of young women volunteered to teach me how to dance, encouraging me to thrust my buttocks farther out and rotate my shoulders farther back. One of the village bands played solely percussion instruments, accompanied by voices of both the band members and the listeners. Another band focused on the vocalizations of the lead singer, who used an electronic megaphone while instrumentalists shook tambourines

Fig. 5.1 A street band performs on drums, aŋkongoma (wooden boxes), and plastic horns in Tor-wama, Bo District, Sierra Leone

made of goat hooves. There were flutes and shakers, drums, an accordion, and makeshift noise makers, dancers in grass skirts, dancers in t-shirts and jeans, one dancer who pulled up her top for effect (Fig. 5.1).

Each village performance was a raucous, joyous, unbridled celebration of Indigenous music that ended with onlookers jumping into the fray. And then there was the crowning musical moment for my companions when they played with the musicians of the Sierra Leone National Dance Troupe under a sprawling tree in Freetown. What a contrast this was, as we sat before the performers for a private audience. In each setting, the music was unique, the setting was unique, the interactions were unique to both the performers and the listeners. And none of the performances, with perhaps the exception of our private audience with the Dance Troupe musicians, resembled in any way a performance that one would find at a Western conservatory or academic music department.

Observing the Observers

This last point is notable, because my colleagues both cite the Dance Troupe experience as particularly meaningful. I believe they took different meanings from the experience, and in my removed position as an observer of the team, I noticed that the setting better aligned with Bernhard's experience as a concert performer than any other in Sierra Leone, allowing him to relax and connect to his own musical familiarity a little more than a village performance would have permitted. We are all creatures of our own experience and making, and that which aligns most closely with our experiences and desires doubtlessly feels more comfortable, thus accommodating our innate predispositions toward expressing ourselves and our skills.

Perhaps this sounds like a statement of the obvious, but if we don't recognize that familiarity produces comfort, which inspires confidence, which engenders authority, we'll be unable to acknowledge how easily we allow ourselves to slide into an authority role by seeking out familiarity. While I'm discussing this in terms of Bernhard's engagement with the Dance Troupe musicians, this is a concept I've thought deeply about while working in unfamiliar—and often uncomfortable—settings as a technical designer. Comfort is a compelling motivator for taking control, and until we can embrace the uncomfortable equally, we risk losing an understanding of place-based knowledge through inadvertent self-confidence.

As I mentioned above, my primary objective throughout our travel in Sierra Leone was to observe how I and my colleagues responded to what we observed, in the hope of identifying implicit biases that could govern our acceptance or disregard of the place-based knowledge we encountered.

This is important, because so many of the world's decision-makers by choice or by default originate in Western society, and while we may delude ourselves into believing our decisions affect no other societies but our own, in a globalized world this cannot be the case. Stop any citizen in the street in the US and ask them who has the most influence in the world, and they're likely to say their own country does (Mennell, 2020). They're also likely to say we have the right of authority over the rest of the world because of our success both economically and socially, in comparison with other countries. This is what Mennell describes as the habitus of our nation, "learned since birth through experience in contact with other human beings, but which has become so deeply ingrained that it feels even to ourselves to be something 'innate' or 'natural.'" Very recently, during travel to the Middle East with several European colleagues, I found myself in a conversation about how marginalized populations regard Western societies. I asked my European colleagues how *they* perceive people from the US. "They're very nice," one of my companions said. "We find them very kind and warm. But they are quite insincere." I'd like to say I was taken aback by this description, but I think it perfectly captures the way we behave outside our own society—we identify ourselves as caretakers of the world assuring the safety and comfort of others, but we also engage in societal blindness when it threatens our self-image of benevolence and, let's be honest, superiority.

If this is, as Mennell suggests, the habitus by which we operate, it most certainly can guide our attitudes toward place-based knowledge associated with technology and music. Bernhard and Jess play an elite brass instrument of great complexity that is present only in European classical compositions. You won't find the horn organically in any Indigenous music, though horns of various types and timbres have infiltrated a number of ethnic genres. The history of the horn is long and glorious, associated with aristocrats and courts of kings. But the African drum has an equally long history and is equally associated with African royalty and aristocracy, an aristocracy of equivalent power and authority within the context of its origins but disregarded in the Eurocentric tradition. This immediately sets up a preordained and powerful bias among elite musicians who have devoted their lives to conserving and promoting a particular sound associated with their own habitus. That bias is difficult to overcome listening to a concert of what appear to be rudimentary instruments carved from bamboo or strung together with plant fiber, particularly when one is accustomed to performing in formal tails on the stage of a grand concert hall before hundreds of silent, adoring listeners.

With my musically untrained ear, I could distinguish complexity of tone, harmony, and rhythm in each of our experiences in both Bolivia and Sierra Leone that rivaled any of the classical pieces I've heard performed in a concert. More than once during our travels, our experiences evoked in my mind the Fauve art movement of the late 19th century, when art masters such as Henri Matisse, Andre Derain, and Georges Braque were disparaged as the "wild beasts" of painting because their works did not conform with standards of the time.

Standards, in fact, play a large role in judging both music and technology as superior, because who sets those standards? Certainly not the marginalized, the powerless, the overlooked and undervalued. In Western society, the keepers of standards in both music and technology are the scholars and gatekeepers of all that is valued. And in both cases, those scholars and gatekeepers come from a tradition that conserves and promotes a particular quality of music and technology that—whether or not it is consumed and consumable by the majority of the world—denotes superiority.

Could Bernhard overcome this habitus of musical superiority during our travels? I believe he did, but it was not without struggle. And it demonstrated to me that when confronted with such a clear difference between Western and Indigenous tradition, one's habitus guides them to favor their own standards and dismiss alternatives as inferior, substandard, or *fauve*.

Jess had a unique vantage point in this Consilience project, being able to explore the concepts of bias and standards both as a musician and engineer. I believe that the diversity of experiences and understanding that comes from the three of us working together— different genders, different generations, different disciplines, and different histories—we were able to identify the implicit biases that each of us deals with when interacting with non-industrialized contexts to recognize how it can limit our own growth and understanding. Expansion of that understanding not only benefits those with whom we work. It benefits *us*, in what we do, in the resources from which we draw, in perceptions of what is right and true and good and quality. And it increases our ability to function across societies by recognizing that whatever we undertake—whether it's orchestral composition or infrastructure design or performance of a horn concerto or development of a new technology—we need to consider the context rather than falling back on our own standards for performance and function.

As is so characteristic of an academic engineer, I have done a splendid job of assessing my colleagues without resorting to examination of my own behaviors and actions. In deference to the self-reflective aspect of Contextual Engineering, I shall now examine my own experiences and biases as well.

I hide behind the mantle of international experience, having worked with people from all walks of life, living in every imaginable setting with a broad range of gifts and talents, limitations and constraints, beliefs and values. But this cloak is merely ornamental, because past experience can only prepare one for their next encounter with the unexpected, rather than assure them that nothing under the sun could be new to them.

In my classes and among my graduate students, I encourage engineering students to think contextually by striving to engage with their client communities using an assimilative perspective (which differs greatly from the type of colonial assimilation to which Bernhard had previously referenced). By assimilating with the client, a Contextual Engineer can better understand the context for which they will design a technology, exclusive of the personal assumptions, biases, and judgments that cloud our understanding of how others view the world. It's a noble objective, but one that I've strived to achieve again and again over my two decades of working internationally, with very limited success. We are all the products of our Western habitus, our instincts and experience—and yes, biases as well. These are always with us as long as we are human, and they will eternally be the voice that whispers in our ear to consider doing things that make sense to us, regardless of whether they align with the client community's context.

And so I'll acknowledge freely that I make missteps, call upon instinct, listen to biases and judgments, whenever interacting with people whose context differs from my own. But to understand the value of place-based knowledge, be it in engineering or music, I must at least recognize the influence of that wicked little whisper in my ear and try to challenge it through observation, communication, and establishment of a relationship of trust with the people for whom I'm designing. It's a difficult task to constantly challenge one's own epistemology, and one that falls in the category of aspirational for those of us who strive to value the contexts we encounter on an equal footing with our own (Fig. 5.2).

Fig. 5.2 Ann sits with women and several men of Santari, Bolivia, in a traditional *pijcho*, during which participants share coca leaves and build a relationship through sharing of knowledge and experiences

References

Mennell, S., 2020. Power, individualism, and collective self perception in the USA. *Historical Social Research/Historische Sozialforschung*, *45*(1) (171), pp. 309–329.
Witmer, A.P., 2020. Contextual Engineering Leverages Local Knowledge to Guide Water System Design. *Journal: American Water Works Association*, *112*(7).

A Bolivian Refreshes Her View of the Country Through Visitors' Perspective

6

Xiomara Echeverria

I consider Bolivia to be a country extremely rich in cultures and traditions in its three geographical regions—the highlands (Altiplano), the valleys, and the plains. Even in our constitution, Bolivia is referred to as a plurinational state made up of original peasant Indigenous peoples and 36 different nations. As I began to travel regularly to small towns and communities—little hidden treasures in Bolivian territory where I could interact with the locals and partake in their traditions and customs—I became curious to learn more about the cultures and customs of Bolivia's Indigenous peoples. Thanks to my work as a Project Manager in *Fundación Ingenieros En Acción*, the Bolivia office of Engineers in Action, I had the opportunity to meet many communities, the vast majority of which were Aymara communities. The Aymara people of Bolivia are found mainly in the Altiplano, where they have a close respectful relationship with their natural environment, a deep reverence for Mother Earth, as well as showing mutual support for each other in jobs such as agriculture, livestock, and community work in general.

Each trip I make has a unique characteristic, and in March 2022 I had the pleasure of traveling to five communities within the municipality of Sica Sica in the department of La Paz accompanied by Dr. Ann Witmer, Professor Bernhard Scully, and two incredible engineers, Jess Mingee and Alexandra Timmons. During this travel, I was able to closely observe the cultural and intellectual exchange that existed between the inhabitants of the communities and the visitors of the University of Illinois Urbana-Champaign. This wonderful team was joined by the director of the La Paz Symphony, Christian Asturizaga, the office assistant of the *Fundación Ingenieros En Acción* Andrei De La Barra, and our

X. Echeverria (✉)
Field Engineering, Fundacion Ingenieros en Accion, La Paz, Bolivia
e-mail: xne2@illinois.edu

© The Author(s), under exclusive license to Springer Nature Switzerland AG 2024
A.-P. Witmer et al. (eds.), *Consilience*, Synthesis Lectures on Engineers,
Technology, & Society, https://doi.org/10.1007/978-3-031-58399-5_6

good and reliable cook Esperanza Condori. Our main objective was to collect information through questionnaires applied to the community members to find out what knowledge they had related to food, energy, and water issues. This was contemplated as a first step of a project that was promoted with the help of the US Embassy in Bolivia and developed together with the Universidad Privada de Bolivia (UPB), *Fundación Ingenieros En Acción* (FIEA) and the University of Illinois. It was our hope that this information could later assist us to create adequate training aimed at leaders and the population in general of not only those five communities but nine others from the same municipality of Sica Sica.

With everyone's involvement, the project was successfully completed and the objectives were largely met. However, none of this would have been possible without that initial journey of gathering information from an engineering and musical artistry point of view. In my experience working with these communities, I was able to learn about their traditions and customs, as well as enjoy their native music. Nevertheless, it was not until the team from Illinois came to visit us to carry out this trip that I was able to appreciate the value of cultural, intellectual, and artistic exchange between Aymara communities and visitors from other countries, as well as witness how people can accept new cultures and musical styles.

The first community we visited was Santari, which had around 380 inhabitants and was divided into two zones. Although some community members became defensive because it was their first time meeting foreigners, the Illinois team and community members were open to the experience, not just sharing information about how the community worked on water, sanitation, and energy issues, but also sharing small customs. I vividly remember how, as Alex, Jess and I began to collect data, Ann was surrounded by the women of the community who were performing the traditional *pijcho,* a practice where people sit in a circle surrounding a traditional fabric called *Aguayo* upon which is placed a wealth of coca leaves. The participants share the leaves, chewing them and holding them in their cheeks until all the flavor and juice is extracted. At the same time, Bernhard and Christian brought out their instruments, a beautiful French horn and a violin that they respectively played (Fig. 6.1), and for that brief moment, the work was so subtly imbued with a cultural exchange that it seemed quite normal for us to be there, witnessing the whole scene in a community setting with the music that Bernhard and Christian gave us. That first day we also toured the community and had the opportunity to speak with local leaders and residents about how the lack of basic needs in the area affects their quality of life.

Checa Belén was the next community, and unlike Santari, this community seemed to have more access to basic needs, and houses were more concentrated and not dispersed as in Santari. Belén had a center square as well as a small tourist center, which was a *balneareo* (place with natural spring pools) that used hot springs for enjoyable communal baths. After collecting the data we needed by visiting the inhabitants' houses, the community members delighted us with an evening musical performance with native wind and

Fig. 6.1 Bernhard and Christian improvise Bolivian and classical tunes on the French horn and violin in the village of Santari

percussion instruments of all sizes. We also found out that Belén has a band of musicians who travel to Peru to play in international Andean music contests, and thanks to the good ear of the musicians in our work team, we were able to enjoy the interpretation of Bolivian rhythms such as *morenada* through instruments such as the French horn and the violin fused to the instruments of the musical band of Belén (Fig. 6.2).

The next community we visited Janko Kollo, a community with a much smaller population and much more dispersed houses. By coincidence, or visit coincided with a visit by the mayor of Sica Sica, which is a major event for a small Aymara community. A musical band from the community played to receive the mayor, and we were invited to the welcome party where we shared food, drinks, a freshly slaughtered goat, and traditional dances. When the community authorities said that the mayor's visit was to listen to a formal request for construction of a community house, we realized that for communities like Janko Kollo, facilities like community centers are considered critical infrastructure.

As we listened to community leaders request funding from the mayor, I was immediately struck by the fact that community members seemed much more concerned about obtaining a two-story meeting house than they were about installing a reliable water distribution system. The village's leader clarified that the money needed to build and operate a water system would far surpass a small community's ability to afford, so they would

Fig. 6.2 A regionally renowned band from Checa Belén plays a Tarkeada, a Bolivian rhythm from the Altiplano

prefer to invest in an infrastructure that could receive people from non-governmental organizations with the hope that water support would come their way through these NGOs. He also mentioned that the village frequently receives visits from NGOs but cannot engage them because they lack the infrastructure to welcome and house them. Our own work team was aware of this since we spent the night there in the house of one of the community authorities (Fig. 6.3).

The people of Janko Kollo get their water from home wells, and in this case, the whole team was able to tour the field and observe the living conditions of the families. All the families had hand-dug wells and used buckets to collect water, but one family had installed a small pump, which was very interesting from an engineering point of view, since, despite the limitations in their education, some inhabitants have knowledge about energies and how to apply them to get the most out of them.

We next visited the municipal seat of Sica Sica, which along with Lahuachaca is the most populated place in the region. Because our visit was on a weekend, and many families who reside there during the week travel to other places, the streets seemed empty. Yet, a local teacher invited us to her house and told us how her building is more than 100 years old. For her, it is very important that places like Sica Sica update information regarding education for primary school children and high school youth, since she feels that they have reached a point where they have stagnated in terms of technology development.

Fig. 6.3 Xiomara talks with the mayor of Sica Sica as he visits Janko Kollo to listen to the requests of residents for a community center

Fig. 6.4 The main square and church of Sica Sica, where the Battle of Aroma took place

Prior to this invitation, some Sica Sica community members were able to enjoy the music of Bernhard and Christian, who began to play in the main square, a historic place where the Battle of Aroma took place, a war conflict of great historical importance in the country. In 1810, a victory was won there, and many historians suggest that it was not only the first victory over the regency of Spain but also the beginning of the Bolivian army (Fig. 6.4).

Fig. 6.5 The Consilience Project team, FIEA staff, and community residents of Culli Culli Alto

As Alex, Jess, and I visited the Sica Sica school to collect more data, we discovered how very different the thinking of one community can be when compared with another, even though both belong to the same municipality. For example, in a small place like Janko Kollo, they were very interested in learning how to disinfect the water whereas in Sica Sica, a peri-urban place, not only did they not seem interested but they were indifferent to this issue. For them, the water that comes from the distribution system is safe, and no resident need worry about anything else associated with the supply.

Our final destination was the community of Culli Culli Alto, which is well known among the tourists who visit Sica Sica for having one of the largest *chullpares* (Aymara cemeteries) reserves in Bolivia. Supplementing the historical site, the community erected a museum in which they housed the group for the night (Fig. 6.5).

Although we were unable to visit individual homes because Culli Culli Alto is a very dispersed farming town, the community members came to the museum to discuss their quality of life associated with water, energy and food. As part of the cultural exchange experience, Bernhard taught one of the community members to play the French horn mouthpiece, and it was wonderful to see how the community members went out of their way to teach Bernhard and Jess to play the native instruments as a thank-you for the French horn lesson one of them received a moment before. It was nice to be part of that experience since I was directly involved as the translator of both parties and in this way, I was able to see Bernhard and Jess learn to play Andean percussion and wind instruments such as the *bombo* and the *tarka* and see to community members of Culli Culli Alto play the French horn mouthpiece. Our final night in the municipality was celebrated inside the museum with dance and music around the remains of Aymara humans and relics of the *Tiwanakota* empire, which dates back almost 3,000 years. The scene sounds bizarre, (and Bernhard was not particularly fond of sleeping in a room with boxes filled with ancient human skulls) but that's the beauty of visiting unexpected places and meeting new people who can alter your outlook on life and inspire you to share new experiences.

The trip I took with Ann, Bernhard, Alexandra, Jess, Christian, Andrei, and Esperanza will always be one of the most memorable, not only because of all the academic information we collected but also because of all the experiences we had in each of the communities. For each new place we visited and enjoyed, for the cultural and historical contributions they gave us, and for the great relationships we forged with each other and with each community, I am grateful. Bolivia is a country so incredibly rich in cultures and traditions that I try to immerse myself as much as possible in the academic, work, and cultural aspects of each trip I make, which leads me to fall more and more in love with my country and its people.

Music is Culture, But is Technology?

7

Philip Foday Yamba Thulla

I have been more involved with Indigenous musicians than I have with Western technology. Most of my experience in Sierra Leone with the Consilience team revolved around music traditions, which is why this chapter contains fewer engineering examples than the many instances of Indigenous music.

Globally, Western technology occupies a privileged elite status, which has shrouded the originality and creativity of Sierra Leone's Indigenous cultural expressions. These expressions have, unfortunately, not been much recognized, leading many Sierra Leoneans to believe they lack the qualities seen in mainstream Western engineering and music. This chapter examines Sierra Leone's Indigenous music and engineering as cultural artifacts and their relationship to Western music and technology.

Customarily, the various ethnic groups of Sierra Leone have kept a united cultural identity, with their music covering a wide range of themes related to cultural moral norms and religious beliefs. The rhythm of the folk music of the Limba people, for example, is strictly based on their social events, which have been used to praise and warn their members. Fertility motifs in performances like *kábɔthɔ* and *åꞒbira* (communal work groups) and demonstrations of filial solidarity in *Bindii bindi mia* (Your brother is your brother, no matter what) are some of the themes explored in Sierra Leone's traditional music. Moreover, the field of traditional music comprises lamentation compositions like Temne *Sampa Soko* (Fig. 7.1), which are often performed during funerals and religious ceremonies. There are also playful and esoteric performances like Temme *Bubu*. These are performed during social or intimate events and they display lively dances and explicit

P. F. Y. Thulla (✉)
Njala University, Bo, Sierra Leone
e-mail: pythulla@njala.edu.sl

© The Author(s), under exclusive license to Springer Nature Switzerland AG 2024   49
A.-P. Witmer et al. (eds.), *Consilience*, Synthesis Lectures on Engineers,
Technology, & Society, https://doi.org/10.1007/978-3-031-58399-5_7

Fig. 7.1 Traditional Temne Sampa Soko Dance in Lungi, Port Loko District, Northern Sierra Leone

music, similar to the Malinke people's *Moribayassa* Dance. Other social events like *lɛnka* and *luknɛ* are exclusively organized by secret society groups such as *Poro*, *Wonde*, and *Bondo*.

Local women and griots (tribal storytellers) of these ethnic groups play a remarkable role in the production of music, songs, dances, and melodies that sum up the Sierra Leonean lifestyle. These compositions unveil activities ranging from agricultural tasks like planting and harvesting to rituals for mourning the dead, and simple moments of enjoyment. This musical variety integrates many instruments such as drumbeats, string instruments, and local percussion to show the rich cultural heritage of Sierra Leone.

Similarly, Sierra Leonean traditional music incorporates many aspects of the surrounding environment. This includes physical, spiritual, and social interactions. This is in contrast to Western engineering principles that focus on problem-solving and the advancement of technology. Engineers in Sierra Leone focus on manufacturing basic tools and objects out of locally available materials. Music, and the oral tradition in general, are usually handed down to 'designated' artisans who are believed to commune with their ancestors, even communicating with the dead. Blacksmiths in Sierra Leone, for example, are thought to be reincarnated and are now residing in communities different from their original communities.

Typically, Indigenous and Western genres have historical, cultural, and social importance. Indigenous music in Sierra Leone performs an even more unique role as a guardian of cultural heritage, with its musicians dedicated to transmitting this music within a specific culture or group, often attributing their artistic inspiration to dreams, alluding to the preservation of their rich heritage.

After the civil conflict in Sierra Leone in the 1990s, the emergence of numerous contemporary instruments compelled most musicians and engineers to accept and incorporate modern technology and engineering. As it was, Indigenous music relied on local materials such as hand claps, clubs, sticks, wild seeds, and tortoise shells, which were integral to daily life and carried profound cultural significance. For example, the bamboo *Bubu* tubes (Fig. 7.2) we observed displayed by the Konike and Mile 91 *Bubu* groups during our visits to Port Loko and Tonkolili were made from materials found in the local environment. However, elements of Western tools—such as car exhaust pipes, carburetors, harvoline rubbers, and tins—were also integrated into the performance to improve sound quality, aligning it with Western musical standards. Many Indigenous musicians found these additions advantageous for tasks like sound mixing and distribution, addressing aspects that were previously considered lacking for a long time. This transformation was mirrored in engineering practices, as traditional huts evolved from thatched roofs, mud, sticks, and wood to structures made of zinc, mud blocks, and boards. This resulted in a process where certain elements of these two art forms shared key characteristics but produced distinct results.

According to Shepler (2010), Sierra Leone's civil war in 1991 spurred local musicians to embrace Western styles of music to address concerns such as violence, corruption, and a lack of job opportunities. Songs such as "Corruption E Do So" by Daddy Saj combine traditional lyrics with Western hip-hop. Modern communication, such as the use of mobile phones, has improved Sierra Leone's music, with many Indigenous musicians using social media to spread their styles and those of notable singers in the past, such as Salia Koroma.

After the civil conflict in Sierra Leone, Indigenous music underwent dramatic changes. Bubu performances in Tonkolili incorporated Western elements such as car exhaust pipes. Kambia musicians used Indigenous and Western instruments such as bullhorns, *akeleŋ* (slit drums), *aŋkongoma* (wooden boxes), and *kondi* (small wooden boxes), to create unique sounds for genres such as *Sampa Soko* and *Disco* music. In the south, Mende musicians used a combination of Western and traditional instruments such as trumpet, *kelei* (slit drum), *shegbureh* (shekere), and *blikuti* (cylindrical drum) for cultural events. This fusion is now seen in the works of prominent Mende musicians like Amie Kallon, Kposowai, Jeneba Koroma, Bobby P, and women's society leaders. These performers, regarded as *numoinasia ti lɔɔ gɔ* (skilled performers), skillfully blend these instruments to amuse their audiences.

Certainly, Indigenous music in Sierra Leone is creative, self-expressive, aesthetic and community-owned. However, community members being the primary creators of their art contradicts the Western idea of copyright. These musicians can improvise and interpret only to convey emotions, cultural values, and stories, and enhance the melody, rhythm, and lyrics of their music. To Western engineers, such approaches emphasize the general concept of shared conceptualization and ownership rather than the "scientific" nature of things. This explains why Indigenous arts are seen as inferior in Sierra Leone.

Fig. 7.2 A Bubu musician displays some hand-crafted bamboo tubes that his ensemble used in a performance in Tonkolili District, Northern Sierra Leone

Fig. 7.3 The *balangi*, a traditional Sierra Leonean instrument, is used in a performance by the National Dance Troupe musicians

The Sierra Leone Dance Troupe, based in Freetown, integrates local instruments with modern tools such as the *balangi* (similar to a xylophone) (Fig. 7.3) and *djembe* drum in revolutionary and traditional *Yeliba* and *Yelimusu* performances. The troupe occasionally performs the *Gorboi* masked devil act and plays string instruments from different ethnic groups such as the Kono, Fula, Kissi, Mende, Vai, and Temne. They also use wind instruments, similar to aerophones, used in secret society rituals, with each ethnic group having its unique instrument for these ceremonies.

In the same way, Sierra Leonean traditional instruments such as slit drums, rattles, and calabashes resemble Western idiophones. The Limba *sambori* and the Temne *aŋrunu* or Mende *jagbawai* drums are styled after European drums. Additionally, local musicians employ improvised tools like saws (*aŋshera*) for unique sounds. *Bubu* musician Janka Nabay digitally recreated *Bubu* rhythms with modern technology, mixing them with hip-hop. A street singer from Moyamba Junction, known as *Man Dɛm* (Men), innovatively uses a rubber sphere as an idiophone. Drizilik's song "Ah dae go dae gbet" (I must go there) is another example of blending local idiophones and Western musical elements (Kamara, 2017). These instruments are played in various cultural contexts. For instance, skin drums (*aŋthabule*) are played before chiefs and their councils, while *aŋsambure* are used in Bondo ceremonies.

The blending of Sierra Leonean tradition with Western concepts of engineering is now quite visible. Multi-story structures and expanded transportation roads are currently being built with support from developed countries such as China. Many locals see this collaboration as a positive step toward greater sustainability and self-sufficiency, but it also suggests that Indigenous structures may struggle to attain sustainability or self-sufficiency without Western assistance.

Sierra Leonean music has grown in terms of its melodic structures. *Disco* music, for example, has evolved to incorporate local chanting with Western rap and hip-hop styles, as well as using distinct attire such as cylindrical hats with cowry shells, and colorful raffia clothes, which has added visual vibrancy to the auditory experience. Instruments like the *kelei* (slit drum) and *sangbei* (drum) now sound like Western lutes and drums. *Bubu* performances use a Western musical scale and numbered sticks for rhythmic variety.

The *Sampa Soko* music blends instruments like *akongoma* drums and *akenke* (a shaker) with chants led by the *Soko*, a spiritual leader. Rituals include rainmaking and cursing. *Kagbɛh* music of the Temne uses traditional and Western instruments, similar to the *Argogo* (instrument with two metal bells) or *Gbatae* (drum) in Mende music, which is accompanied by singers and dancers.

Despite being stylistically similar to Western genres, Sierra Leonean music often explores cultural and political themes. Because of its roots in the Temne secret societies, the traditional *Sampa Soko*, for example, involves ritual chants and regimental steps for initiates. *Kagbɛh* music contains political content and is played at social events for educational purposes. It also promotes or satirizes local leaders (Fig. 7.4). *Bubu* music, while mainly performed for entertainment, can occasionally spark competition and provocative performances, particularly during Sierra Leone's Independence Day celebrations. These musical expressions, which stem from a desire for sustainable communal life, link social needs to available resources. This is seen in the local musicians' use of Western instruments like pot lids and nylon strings in traditional genres, which aligns with Western philosophies on production and artistic control.

Drawing on my experience working with traditional musicians, particularly my participation in The Consilience Project in December 2022, I recognized that there were still challenges in blending traditional music with Western styles. The locals expressed sadly that their Indigenous music was dying. They also expressed that there was little opportunity for Indigenous music to gain widespread recognition or be compared with Western standards. In addition, factors such as nationality and cultural differences frequently hampered the effort to digitize and adapt Sierra Leone's Indigenous music for Western audiences.

Even if planning is necessary for Sierra Leonean artistic creativity, spontaneity is seen as an important component of Indigenous music. This sets it apart from the more prescriptive approach in Western music production. Many technical and design innovations encourage spontaneity, with little planning required for ideas. However, concerns have been raised regarding the quality of culturally manufactured goods such as aluminum pots,

Fig. 7.4 Orduley and His Kagbɛh Traditional Music Group perform in Kambia, Northern Sierra Leone, using a modern bullhorn to enhance his vocalizations

cooking spoons, and brick-making machines. Finnegan (2017), for example, has expressed concern about the digitization of oral practices, claiming that doing so compromises the key characteristic of word-of-mouth transmission. This is in contrast to the Western perspective, which sees digital technology as increasing output and making information easier to access, store, maintain, and share.

Throughout history, Indigenous tribes have transmitted knowledge and skills through oral traditions and apprenticeships, doing so mostly through live performances that allow the audience to share a cultural experience. Locals in the communities visited describe a lack of understanding of traditional infrastructure and design, as shared cultural knowledge rapidly declines. This could be the reason why Sierra Leonean engineers use Western engineering techniques and seek collaboration for infrastructure development. A similar reliance may be seen in music, with current musicians using mobile apps to improve their performance.

A sad truth is that Indigenous music in Sierra Leone has been ignored by its creators due to modern technology. During our last visits, we had discussions about the equipment used by the musicians in their performances, and they expressed the desire to Westernize Indigenous music and its instruments. Many of the locals viewed creating and transmitting Indigenous music as a shared art form that had been passed down from generation to generation. The younger generation showed a preference for Western-styled music like hip-hop and dancehall, as well as the demand for cell phones, computers, and new infrastructure. This may justify the claim that Indigenous arts are inferior to mainstream or Western ones.

However, a positive shift is emerging as global recognition and respect for Indigenous arts grow, despite the persistent threats posed by the mining industry and large-scale infrastructure to local heritage. Traditional musicians have adeptly crafted and conveyed complex musical forms through their instruments and compositions. My observations during our visits reveal that, despite their similarities to Western music, traditional musical instruments and performances like *Bubu* and *Disco* will always be community-owned and shared. Nonetheless, Indigenous musicians have developed ways, such as using multimedia, to sell their content similar to Western conventions, which has helped to revitalize Indigenous performances.

As a result, Indigenous music should not be seen as lacking in skills simply because it is rooted in cultural expression, creativity, and preservation of cultural heritage, as opposed to Western technology, which is based on scientific principles, but rather as a mixed bag of artistic creations that embodies the essence of the civilization that crafted it. This necessitates a comparison of music and engineering, which allows for a more in-depth understanding of the fundamental aesthetic notions in both creative forms, as well as the problem-solving approaches that span these seemingly unrelated professions while respecting their cultural values.

References

Finnegan, R. (2017). Oral literature in Africa (p. 614). Open Book Publishers.

Kamara, E. (2017). Bubu in Sierra Leone. Music In Africa.

Shepler, S. (2010). Youth music and politics in post-war Sierra Leone. The Journal of Modern African Studies, 48(4), 627–642.

Part III
The Consilience of Music and Technology

The Common Thread of Engineering and Music

8

Alexandra Timmons

"Music is the universal language" may be a sentiment you have heard before. Like spoken word, dance, art, and other forms of self—and community—expression, nearly, if not all, human cultures across space and time have developed music they call their own. Now, referring to music as a universal language is not to be confused with universal comprehension. The use of music across cultures serves a vast array of purposes from expression of happiness to sorrow, consists of a variety of harmonies from soloists to extensive orchestras, and presents itself in a range of ways from free form to strictly structured. The various elements that compose the notes and sounds we hear are a complex study on their own. But the heart of music is consistent no matter where you look—a rich form of human expression that prevails even in times when words cannot.

On the other hand, engineering had always stood out to me as rigid and objective. This rigidity was something that was reinforced in my undergraduate engineering courses. There was always a singular "best answer". If you are designing a water supply system, to get water from A to B, there are optimal materials with which to construct, an optimal path to take, and an optimal end result. And this way of thinking can work just fine when you are an engineer working in areas very similar to your own. The materials you have access to and the laws and regulations you are operating under, for example, are likely nearly identical. But what happens when we step outside of the familiar?

In my experience, we tend to lean into our rigid training even further. If there is always a "best answer", what benefit is there to learn about elements of a community such as its culture or political landscape? Isn't all that should matter the objective data

A. Timmons (✉)
Illinois Applied Research Institute, Champaign, IL, USA
e-mail: atimmon2@illinois.edu

that gets us to that familiar "best answer"? Engineers who open themselves up to the world of Contextual Engineering (CE) know this is not the case. However, when the CE methodology is new, all the various components that create a community's unique identity swirl in your head, and you can easily find yourself overwhelmed. How does one fit all of this new and unfamiliar data into the structured formulas and methods we learned in school? And on the flip side, what is this community supposed to make of this group of engineering "tourists" who may not even speak the same language as them? What reason have the engineers provided to gain their trust? Of course one could take Ann Witmer's immersive CE course or read her comprehensive textbook to uncover answers to these questions, but this book proposes an additional avenue—music.

What I find so striking about music is that there is always a thread of familiarity. I may empathize with the emotions expressed through the melody or recognize a familiar and steady drum beat. Even in music that is wholly unlike what I am used to, there is always something there to which I can find myself relating. In a world where humans are quick to notice differences, when we are finding ourselves overwhelmed working in an unfamiliar space with unfamiliar data, this thread of familiarity can pull us back together.

In March 2022 I had the good fortune to travel to Bolivia to help develop a sustainable technology training program. This program development was a partnership between three engineering entities: two Bolivian organizations, engineering non-profit *Fundación Ingenieros en Acción* (FIEA) and academic institution *Universidad Privada Boliviana* (UPB), and our team of CE researchers from the University of Illinois. The intent of our visit was to travel to different rural communities throughout the municipality of Sica Sica that would partake in this training program to generate a baseline understanding of their current needs, capabilities, and limitations with regards to food, energy, and water. Travel of this nature was not unfamiliar to us in CE as we have conducted similar contextual investigations in countries all over the world. However, what made this trip different was the composition of the travel team. Joining our cohort of engineers was a renowned American professor and French hornist, Bernhard Scully, and acclaimed Bolivian concertmaster and violinist, Christian Asturizaga, as well as the brilliantly talented dual engineer-musician, Jess Mingee.

The inclusion of musicians on this trip was not driven by a specific hypothesis or bias intending to benefit the technology training program we were there to begin developing. It was just based on a simple idea surmised between an engineer and a musician that the two vocations may have more in common than we give them credit for. In fact, our team initially considered the projects two entirely separate efforts, one was the technology training program and the other informally comparing and contrasting music and technology. But on that first day in Sica Sica, when just the sound of music was enough to enhance the relationships and information sharing between the engineers and the community, the projects no longer seemed so distinct.

Fig. 8.1 Christian freestyles on violin while Ann dances with a resident in Santari, Bolivia

In front of the Santari community center, free-styling horn and violin strains accompanied our conversations with residents about their food, energy, and water needs. Bernhard and Christian had pulled out their instruments and began making music, pulling elements from each of their cultural and professional backgrounds and melding them together in this new environment. More and more residents began to gather around, drawn by the music. Drinks were brought out to share, and shortly after people began dancing, even inviting our researchers up to dance together with them. While we as foreign researchers typically find ourselves on a slow but steady path to mutual comfort in these sorts of scenarios, the introduction of music seemed to have helped bypass that initial hesitancy period. It was almost as if we all shared a release of nerves and formalities and instead found ourselves laughing and interacting naturally, even without a common language (Fig. 8.1).

Throughout our time in the rest of our partnering communities, we were fortunate to share similarly meaningful experiences. In Checa Belén, for example, the engineers focused on collecting data in the morning, going door to door looking to speak with anyone who was willing. It seemed we had covered every square inch of the community and not yet met our ideal quota of interviews. However, as soon music flowed from Bernhard, Jess, and Christian, an impromptu evening concert seemed to effortlessly pull people out like a magnet. Nearly a quarter of the community stopped by over the course of the night to spectate, dance, and even participate, bringing instruments of their own and performing for the crowd. The music made by our team was wildly different from the music made by the residents. The expressed tonalities, rhythms, forms, and styles were clearly stemming from two different cultures. But the ambition of both groups was the same, to perform, share, and experience the joy of music together. This was the common thread.

Welcoming guests into a community can be a cautionary adventure, especially when there are existing stereotypes dividing the groups such as the perceived assumptions that come along with the titles of "Western" and "Non-Western". These hesitations can be quickly exacerbated when the topic focuses on something like technology, which the west tends to view as a linearly progressive subject matter—if you are not seeking out the latest and greatest technological advancements, you are behind. In my experience here in the United States, there is an air of superiority that tends to follow those who possess the newest that technology has to offer. This attitude is ingrained in our society, and thus in our educational system. We are taught about the wonders of modern advancement and the incredible things that we are afforded as a result of those innovations. We tend to then carry through this mindset into our work as engineers, pushing the perceived "best" on the clients we work with. However, as my work with the Contextual Engineering approach has taught me, not everyone sees things the way we as engineers tend to in the United States. There is never one singular right answer to a problem such as growing a community's food, energy, and water self-sufficiency and sustainability. The unique makeup of each group of people united by some shared identity necessitates unique solutions. If the way a community will use and interact with your design will vary, shouldn't that dictate that the design itself should vary in response? How then can we expect to adequately adapt designs to a community if the best perceived solution is always simply the newest (Fig. 8.2)?

Fig. 8.2 A domed structure, erected by a Janko Kollo landowner to protect his equipment, is a common sight on the Altiplano

In Janko Kollo we were introduced to a building that was not immediately recognizable to our team. It was a small structure, cylindrical in shape, tapering inwards, with a domed roof on top, and an opening large enough to fit a person but without a door to enclose it. The entire structure was composed of organic material including earth and straw. Ignorant of its use or purpose, my initial thought was that it appeared temporary, likely did not house anything of value since there was no door to secure its contents, and probably was difficult to construct but done so out of necessity. With this initial thought, my brain left no room to find the common thread. I immediately recognized the differences, and completely ignored any chance of finding similarities. Had I not explored further, I would have walked away from this feat of construction with a totally incorrect point of view.

This viewpoint would continue to consciously or subconsciously affect my opinion of what would be the ideal technology training program design for this particular community. But fortunately, we had shared a meal and some music with our guide prior, and even though this structure was not a part of our initial exploration I did not hesitate to inquire about it. As you may have guessed, the truth about it is nearly entirely different from my initial thought. We learned how the particular mixture of earth and straw keeps the building cool in the heat and preserves warmth in the winters. We learned that since the structure is built from the land, it is the most environmentally friendly construction option available, and is strong enough to withstand the elements of the area. These structures could be built in incredibly short periods of time with minimal manpower. These design benefits are blindingly similar to those we seek in Western construction. This one simple conversation illuminated not just a thread, but a whole spool's worth of familiarities and opportunities for connection. The context of this structure offers critical insight into a community's values, resources, and priorities but is completely missed when we are not prepared to reject our immediate disposition to recognize differences in favor of seeking information that instead brings us closer together. It struck me how, had we not connected the night before as musicians in the community center, we might not have shared this uniting experience as engineers out in the field.

While it's not feasible to have a traveling band of eminent musicians accompany engineers on every trip, elements of this experience can certainly be replicated to the benefit of future projects. For the average person, discussing technological developments can be overwhelming and at times frustrating, if the person even feels "qualified" to be discussing them at all. But discussing music is a passion, a great icebreaker conversation topic, and frequently, a gateway to common ground and even friendship. Allowing time to connect with the people you are working with opens the door to revelations and design ideas you may have never come across, ultimately serving the long-term sustainability of the project itself. And music is a great avenue to start with. Similarly, entering a new experience with an openness to not just teach, but be taught, whether it's music or technology or just about anything else, can make a world of difference in helping to break down the societal barriers that keep us from recognizing the commonalities between us. That thread that I had always felt connected musicians the world over was not limited to bringing people together in a musical space. In fact, that thread could connect people across all sorts of perceived boundaries, including the seemingly rigid world of engineering. Now, as a working engineer, whenever I find myself struggling to understand the heart of a problem or the perspectives of the people most closely impacted by the issue at hand, I remind myself to search for that thread. If we can find that common thread that can turn a total stranger into a friend, imagine how we could elevate our work as engineers and beyond (Fig. 8.3).

Fig. 8.3 A quena (Andean flute) player performing at the Valle de la Luna (Moon Valley), a nature park near La Paz, Bolivia

Using Consilience as a Bridge Between Music and Technology

9

Ariel Urquidi

I would like to share how The Consilience Project has impacted my perspective on music and engineering. Ann Witmer introduced me to this concept while working on a Contextual Engineering class for a global classroom project for our two universities. We exchanged ideas and discussed how this knowledge could enrich our understanding of these subjects.

I began to see the connections between music and engineering more clearly. Exploring how these seemingly disparate fields could inform and enhance each other was fascinating. I have become more interested in exploring the intersections between different areas of knowledge and how they can be applied in new and innovative ways. By studying the intersection of these two fields, we can gain a deeper appreciation for civil engineering and music's technical and artistic aspects.

Music has been a part of my life since childhood. I was exposed to wonderful music from a young age and at the age of 10, I began taking flute lessons with aspirations of becoming a violinist. However, my teacher suggested that I would excel as a drummer, and that's what I pursued. I've been playing for nearly two decades, but in the last five years, I've taken my practice more seriously. I establish a daily routine, warm up properly, and study diligently. The experiences I had in developing my skills as a progressive rock drummer actually mirrored my development as a civil engineer.

During my school days, I had a strong fascination for Physics and Mathematics. It prompted me to pursue an engineering major that would align with my skills. I initially enrolled in Electromechanical Engineering, but eventually shifted my focus to Civil

A. Urquidi (✉)
Universidad Privada Boliviana, La Paz, Bolivia
e-mail: arielurquidi@upb.edu

Engineering. Today, my Ph.D. in environmental engineering and water management is pending.

It is interesting to note that we don't always have to learn in a scientific way but can learn from others simply by observing or listening. During my early days of learning music, I relied on watching videos, listening to music, and air drumming along with them.

During my time as a junior Civil Engineer, I had the privilege of working as a supervisor in a soil mechanic's lab and learning from an exceptional technician named Freddy. Although he did not hold any professional title, he had gained valuable knowledge from another technician. On the first day of the lab, I arrived with my Braja M. Das book, feeling confident that I knew everything because of my completed degree. However, I quickly realized how wrong I was. Initially, I noticed that the sample preparation techniques we used in the lab were different from what I had learned from the book. I pointed out to Freddy that it was not the correct way, but he just smiled and calmly replied, "We will do it your way and our way." It turned out that their method was much better than mine.

Learning about real life is beautiful, as it is not always based on theory or books. I always advise my students to balance empirical and theoretical learning.

A consilient lesson from these examples is that in both music and engineering, we learn by doing, by making mistakes, and by listening to others. Rather than assuming an expertise or knowledge, as I had in the lab, I needed to approach my technical work in a way comparable to my musical development. I needed to practice, practice, practice.

Now that I know what consilience is, how can I put it to work for me? A lesson from consilience could certainly come into play, for example, when I'm creating a new progressive rock song. It's important to keep in mind that in both engineering and music, not everyone will be impressed by what you create. While your songs may be great for some people, others may not enjoy them as much. Similarly, in engineering, your solutions may be creative and effective, but they may not be well-liked or well-received by your colleagues or users. Accepting and understanding that not everyone will have the same taste or appreciation for your work is crucial.

Let's look at a unique architectural example in the City of El Alto, Bolivia, for another example of consilience. The *cholets* are essentially buildings constructed on top of other buildings. While they may not be aesthetically pleasing to everyone, the owners of these buildings are some of the wealthiest people in El Alto. They use these buildings to run their businesses, renting out the larger salons for parties and the small shops in the front. Some even live in the rooftop apartments of these buildings. Despite their unconventional appearance, these *cholets* serve a functional purpose and have become a symbol of success for their owners.

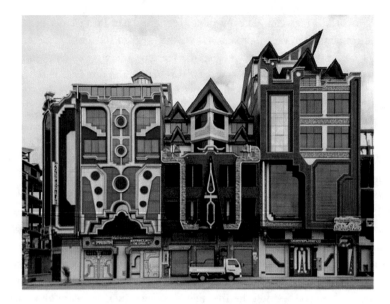

There's an odd harmony of purpose, function, and appearance in these buildings, and there's also an incorporation of strong technical principles that permits the structural components to support, literally, the architect's grand vision.

The blending of principles and vision is no different for a songwriter, who draws upon their knowledge of music theory, history, and culture to create a technically innovative and socially relevant song. Neither component stands alone, if the objective is to create a song that doesn't offend both the senses and the sentiments of the listener. Ultimately, the goal would be to create a song that not only sounds great but also resonates with listeners on multiple levels, drawing on different areas of knowledge to create something truly unique and memorable. Contextual engineering fulfills this same purpose by addressing user needs and listening to the users.

Sometimes, musicians collaborate with other artists and experts to refine their ideas and ensure their lyrics are as impactful as possible. Ultimately, the goal would be to tell a compelling story and connect with listeners on a deeper level. As engineers, we read various researchers and scholars and analyze different methods to acquire expertise in different fields. However, it's equally important to acknowledge that listening to our colleagues, talking with them, or joining them on their field trips can enhance our knowledge and wisdom. Maybe we can even explore different approaches to engineering an infrastructure or technology. This could be considered comparable to a musician exploring a new genre of music.

Regarding music and engineering, it's crucial to stay informed and inspired. Challenging ourselves is key, as it's easy to become burnt out or run out of ideas when composing a new piece or solving a problem. When it comes to engineering a solution, I have sometimes struggled with math or calculations. It's not unusual for me to find that listening to one of my favorite songs can ease my mind, relax me, and bring inner peace. I know it may sound romantic, but this approach has always worked for me.

When it comes to drumming, whether it's learning a song or composing something new, I always take an engineering approach. I remain calm, analyze if the solution that presents itself most immediately is the most suitable, and always have a backup plan, sometimes even a plan C or D that combines the previous ones.

Based on my personal experience, I can strongly relate to everything I have previously described, especially with the recent developments in creating a new album with my band, Innerwake. As a musician with a background in engineering, I have come to appreciate the significance of organization, scheduling, and leadership during rehearsals and music creation. These aspects are crucial to my overall success as a musician. I often feel as though I am working on another engineering project as I continually design, redesign, and ideate ways to implement superior solutions. It is a challenging and rewarding process that allows me to apply my skills and creativity in a unique way.

Nowadays, I am working with my band on the production of our forthcoming album called "The Phases of…", and we are proud to have created a concept album that explores the consilience concept through the five phases of the duel. We carefully created every song by drawing on various knowledge and life experiences to ensure they are both intellectually engaging and emotionally impactful. Our main goal was to produce an album that not only delivers excellent sound quality but also narrates a captivating story that resonates with the audience on a profound level. We dedicated all our creative resources to crafting something exceptional and significant to achieve this. Using different musical styles and tones to convey each phase, the album will provide a unique and immersive experience that captures the complexity and drama of a duel.

Much like an engineering design, we could have approached this album by focusing on the first musical approach that presented itself, regardless of whether it captured the full mood and feel of the track. In the course Ann and I teach together, we call that tactic *satisficing*, or settling for the first available option that presents itself as satisfactory. The risk of satisficing in both music and engineering is that it prevents the creator from finding the *best* option. In the case of our album, a satisficed approach likely would have yielded songs that were acceptable but repetitive. In the case of an engineering design, satisficing is likely to yield a technology that works, at least in the present, but doesn't necessarily fully address the needs and desires of the users.

It's worth noting that my band plans to incorporate the roots of our Indigenous ancestors into our music in the future, as we have a diverse and vibrant culture. In my opinion, Bolivian music is a real pleasure, but it may take some time to appreciate it fully. You should give it a chance, even if the high-pitched voices or folk style may not be your initial preference. It's all about taking the time to embrace it and allowing yourself to enjoy it. It encompasses a range of genres, from fast-paced 16th-note tunes to romantic ballads that resemble waltzes. The most iconic of our national music is known as *cueca*, where you clap every two or three beats to keep up with the 6/8 rhythm while dancing and singing.

Consilience has the potential to revolutionize our experience and approach to both music and engineering. It can be applied in many ways, such as using old tires to create innovative furniture or to stabilize natural slope formations. Additionally, it can be used to infuse your favorite riff from a rock song into a new composition, creating something entirely new and unique, also by leveraging technology in innovative ways, we can push the boundaries of what is possible and create new forms of artistic expression. From designing cutting-edge instruments to developing new software that enhances the creative process, there are countless opportunities for engineers and musicians to collaborate and make a lasting impact on the world. Whether creating new music genres or developing advanced sound systems that deliver unparalleled audio quality, the possibilities are endless when music and engineering come together.

How Can We Challenge the Status Quo with the Ground Truths We've Found?

10

Jess Mingee

Sierra Leone Improvisation

It was during a particular experience in Sierra Leone that I came to understand the full gravity of consilience. Our time in the country was mostly focused on the exploration of music, and we had the opportunity to observe and interact with several groups from different communities. One day in Freetown, we met with the National Dance Troupe, a well-known group of performers including vocalists, instrumentalists, and dancers. Although we did not meet any dancers, a handful of instrumentalists and vocalists engaged with us. We first listened to them perform on their own, demonstrating a variety of traditional and popular music. Afterwards, they were curious to hear some of our music and play with us, so we decided to attempt a musical improvisation together.

They asked us to play something on our horns so they could get an idea of the sound. I was expecting we'd play for a brief moment and then we would stop and talk through a loose plan for how we were going to play together—to my surprise, the rest of them jumped in within seconds of us, providing their own baseline beat. The resulting sound was not very cohesive because their traditional beat was at odds with our swing/blues style. But gradually, we adjusted the baseline melody on our horns to match more closely to their beat. Then, a vocalist joined in with a complementary melody and before we knew it, people were naturally falling in and out of solo roles as they exchanged ideas with each other. By the end, we had gravitated to a harmonious collective sound because Bernhard Scully and I adapted to their music and supported them with our horns. It was

J. Mingee (✉)
University of Illinois Urbana-Champaign, Urbana, IL, USA
e-mail: jmingee2@illinois.edu

Fig. 10.1 Collective improvisation with the National Dance Troupe in Freetown, Sierra Leone

so exciting for me to be able to experience music in this way and converge towards a coherent improvisation with people and instruments I had never met before (Fig. 10.1).

I was blown away by this experience, and it was a moment of clarity for me in how my understanding of music could influence my understanding of engineering and vice versa. While each of us first jumped in with styles representative of our personal musical background, our act of diverging from those styles led to the cohesive musical synergy we achieved with the National Dance Troupe. The various adaptive skills required of us in that improvisation are reminiscent of the strategies I have learned in Contextual Engineering when working with new and unfamiliar user populations, since both call for the following:

- Awareness of power dynamics: we did not try to assert our own expertise and instead had the humility to make space for input from the Troupe
- Readiness to pivot from our initial vision: Bernhard and I started with a particular baseline but we adjusted it after hearing what others were playing
- Willingness to go "off script" from a melody/technique that is familiar to us: we attempted to echo some of their tunes or rhythms despite being unaccustomed to them

While I had previously seen my studies in engineering and music as separate, this exploration opened my mind to the world of comparison in these two fields. My understanding of the relationship between Western and Indigenous knowledge has strengthened by looking at how it presents across these disciplines. I have grown as both an engineer and a musician by realizing that the wall I had put between the two was unnecessary.

Parallel Stories

I wasn't always able to improvise music, let alone with people and instruments from another country that I am visiting for the first time. I felt perfectly comfortable and happy reading music from a page and being told exactly what to do, and I had no interest in trying to improvise music. This was largely influenced by my musical upbringing and how I was trained. Akin to Bernhard's experience, I was conditioned towards classical music from a very young age. I took piano lessons starting in kindergarten which gave me an introduction to music theory and rhythm. In sixth grade, I switched from piano to French horn, and I was very quickly attracted to the structure it provided. I learned how to position my hands and lips, and how to read music in the key of that instrument. Every rehearsal was the same, completely predictable, and that is one thing I enjoyed about it. There was a simple but unspoken rule that if I learned certain skills in certain ways, then I would be successful in music. I blindly followed that path all the way through high school, meeting most of my musical goals because I did what was expected. When I went off to college, I looked forward to more years of exactly the same thing.

In one of my first lessons with Bernhard in the fall of 2018, he asked me to play something improvised. As he mentions in his essay, my immediate response was something along the lines of "I can't." From my limited perspective, improvising was something jazz players did, and I was not a jazz player. However, I would reluctantly attempt to improvise off and on when asked. In my earlier attempts, I would ask misguided questions like "Did I do it right?" or "Is that how it was supposed to sound?" Additionally, I would secretly practice what I would "improvise" before my lessons to avoid actually improvising in them. I was uncomfortable with improvisation and did not understand why I needed to be doing it.

My affinity for the structure I had in music was a big part of why I decided I wanted to study engineering. I chose it because I liked the clear-cut nature that I perceived it to have. As Alex Timmons discusses in her essay, we were taught that there would be a single "right" answer to be found as long as we followed the correct procedure. Much like my early experience in music, math and science were predictable and there was a prescribed formula to achieve the societal definition of success. However, another passion of mine within engineering was the desire to help people. Specifically, I was interested in Engineers Without Borders and their mission to utilize engineering for the benefit of humanity globally. I headed to college ready to learn whatever formulas I needed in order to help people in a higher-purpose kind of way.

Similar to my experience in music, the fall of 2018 would challenge my preconceived notions of the predictability and certainty of this field. In my first semester I attended a lecture by Ann Witmer about something called "Contextual Engineering," which would turn out to be my wake-up call about what effective engineering *really* means. I realized I could not follow a formula to address complex societal problems with engineering, nor could I identify one solution to a problem and distribute it everywhere. I learned that I had to be willing to adapt my engineering approach to a given context, which would come with its own uncertainty. Throughout my undergraduate years, I worked to obtain not only technical engineering skills, but the ability to uncover and assess all the factors surrounding a problem and how they might impact design. I grew to understand that there was a complicated relationship between "more-developed countries" and "less-developed countries" and that the power dynamics created between Western engineers and non-industrialized communities have an unavoidable impact on stakeholder relationships and project decision-making. Furthermore, I observed a tendency of Western engineers and engineering curricula to prioritize our own knowledge over the local or Indigenous knowledge present in client communities, based on a perception that those methods lack merit or proper scientific foundation. I heard multiple stories about how ignorance of local knowledge and techniques ultimately led to inappropriate or failed technology.

While my perspective had evolved in engineering, I still viewed music more or less the same. I viewed it as an outlet, a time to do something completely different from engineering, and I considered it a separate world. The issues I had come to know in engineering had nothing to do with music…until one of my lessons with Bernhard in my second year where I heard him talk about the positioning of Western classical music in the department and how he was told it took priority over all other kinds of music on the basis that other kinds are "not really music." I was immediately reminded of some of my discussions with Ann where she shared how she had been told that utilizing Indigenous knowledge or tools in engineering projects wasn't "real engineering." I decided to put the two of them in touch, not realizing at the time what potential this crossover truly had.

Breaking the Divide Through Improvisation

Our visit to Bolivia was the first opportunity I had for direct exploration of the relationship between these two disciplines in an international context. As we travelled to several different rural communities and interacted with them from both an engineering and music perspective, I gained many insights. When it came to engineering discussions, I was fully equipped to think and interact across cultural boundaries. I remember I was able to confidently ask different residents about their lived experience and perceptions of technology. When it came to music, however, I felt empty-handed. We met with Indigenous musicians and had the opportunity to play our horns for them, and I completely froze. I did not even attempt to improvise, and I was only comfortable playing the written music that we had rehearsed before the trip. I then understood how music improvisation would be an important and useful skill for me to learn.

The ability to improvise without a preconceived plan was a skill I had cultivated in engineering, but not in music. Seeing how important it had been in engineering motivated me to pursue it in music, daunting as it was. Despite an initial aversion to the idea, in the fall of 2022 I decided to join the Improvisers Exchange Ensemble led by Jason Finkelman, who directs the Global Arts Performance Initiatives on campus. To my surprise, the class completely changed how I view and approach music-making in a positive way. The ensemble welcomes students of any musical background and experience level and puts no restrictions on instruments or even what defines a musical instrument. It was that kind of environment that taught me to be comfortable engaging with music that wasn't pre-decided or written on a page. The only way I could have engaged with the variety of instruments and people in my class was to learn to tolerate venturing away from the classical music training that I had been given for years. I had to learn to adapt and apply that knowledge in a way that fit in with the other sounds I was hearing. In many ways, this ensemble reflected the Bolivian approach to music that Christian Asturizaga describes—we all had to be willing to think that seemingly unrelated sounds could blend together towards one musical product. Instruments such as *oud*, *berimbau*, French horn, *hulusi*, synthesizer, and a balloon suddenly had to find a way to play with each other.

This is a huge parallel with engineering: Contextual Engineering has taught me that in order to make an appropriate design in different contexts, we have to be willing to adapt our design process to the particular conditions we are facing. It is not as simple as following a formula. Once I realized that connection between engineering and music, I was able to grow in my improvisational abilities, understanding that it was the most effective way to be able to play music in other settings besides classical band or orchestra. While I still may be more comfortable reading music off of a page, or following a pre-determined design procedure, my willingness to depart from that structure in engineering shed light on how I also needed a broader and more flexible approach in music. In other words, my Contextual Engineering training had given me leverage to learn how to improvise music. My newfound improvisational abilities allowed me to have the experience that I had in Sierra Leone, which was my epiphany moment towards the concurrence of these two fields. Through that experience, I noticed how skills I had already cultivated in music could now give me leverage in my approach to engineering. The ability to listen others in real time and use what I am hearing to formulate my next move, for example, would also be useful in an engineering collaboration. There are even more parallels than I originally thought between engineering and music with regard to cross-cultural connections: perceptions of sophistication or merit, power dynamics, and structure in academia.

Perceptions of Sophistication or Merit

Since I was only comfortable with my own knowledge and experience in music while visiting Bolivia, my interactions with local musicians were limited to what I could observe through Bernhard and Christian. This mindset also manifested while I listened to local musicians perform; I had a difficult time appreciating the music simply because it was different from my own experience. In fact, many characteristics of their performance

Fig. 10.2 A local moseño ensemble performs in Checa Belén, Sica Sica

such as tone quality or coordination were the opposite of what I had been taught in my music education. It was loud and overwhelming, and the tone coming from the wind instruments sounded overblown. I didn't particularly like listening to it, and I realized that is because I was comparing it to my own standards for what defines a desirable sound. I later learned from a local Bolivian musician that louder, more forceful sounds are preferred and seen as bringing life as opposed to soft, pretty sounds which paint an image of hunger, or needing more. As Christian articulates in his chapter, the very sounds that Western music considers to represent "tension" are actually considered relaxing in Bolivian tradition. Once I adjusted my mindset away from my own standards, I began to really appreciate and enjoy the different performances that local musicians so kindly shared with us (Fig. 10.2).

I was able to draw this conclusion and adjust quickly because in engineering, I had already learned that Indigenous knowledge could prove to be just as effective, if not moreso, than Western technical knowledge when designing infrastructure in certain settings. Furthermore, there might be a specific use or cultural significance to certain tools or techniques that is important to acknowledge even if we do not understand it. In his essay, Ariel mentions how the *cholets* in El Alto provide both useful function and positive reputation for business owners despite having an unusual appearance compared to traditional infrastructure. Xiomara mentions in her essay that Janko Kollo decided to spend money on a two-story community center instead of water distribution infrastructure, which seems counter-intuitive to the visiting eye. However, we discovered that the community found it more effective to invest toward a space that can welcome outside groups for building water infrastructure rather than directly fund the infrastructure themselves.

Another way this bias materializes is through the Western-minded tendency to assume that communities in "less-developed countries" all have similar music, or function similarly enough that one engineering solution can be replicated broadly. Our travel experiences challenged this notion when we visited several communities in proximity, only to discover that they were quite different from each other. In Sierra Leone, we would barely drive for fifteen minutes before reaching a community that gave a completely different performance in terms of rhythms, instruments, attire, and social customs. For example, in one community, only certain people were allowed to be musicians and we sat down to watch them perform a curated series of songs. Later that day in a different area, we danced along with a large gathering of community members while musicians performed more informal music that involved a call and response structure with the audience. As Xiomara illustrates in her essay, we had an analogous experience in Bolivia, visiting communities right next to each other only to find that they had different solutions to similar engineering problems and differing levels of acceptance with certain technologies. Where one community gravitated towards bleach as a water treatment method, another community actively opposed it, explaining that bleach would ruin the sweet taste their natural water has.

In addition to the diversity and cultural significance noted above, the complexity and value of Indigenous traditions could easily be overlooked. One of the performances we watched in Sierra Leone switched songs several times with perfect coordination among everyone, and I sat there wondering how it was done so precisely without any kind of conductor or obvious visual/audio cues. I later learned from one of the musicians that this was accomplished by the dancer making eye contact with the leading drummer. In another community, we learned that the musicians are essentially the moral authorities, in charge of settling disputes, which was something very unfamiliar to me. In terms of instruments, one of the Sierra Leonean communities in the northern Temne region performed with a ten-part system of *Bubu* tube instruments skillfully crafted from local bamboo; as Philip mentions in his essay, those were completely unique to that area and hold significance by representing the community's connection to their physical environment (Fig. 10.3).

From the engineering perspective, in Bolivia I observed creative solutions such as using hollowed-out car tires as a means of retrieving and treating water, or maintaining more than one property in a community such that cattle can always have grazing space while a separate area is regrowing. As other contributors have mentioned, we also had the opportunity in Bolivia to visit an ancient Aymara burial site with tombs (*chullpares*) constructed from adobe which are still standing after over one thousand years of weathering. Current efforts to recreate adobe of comparable strength have all failed, as the process behind the original construction was lost.

Fig. 10.3 A local musician explains the significance of the Bubu instruments and why they are unique to his community in the Tonkolili District, Sierra Leone

Despite years of music training and an engineering degree, I do not have the diversity of knowledge or thought processes that the various Bolivian and Sierra Leonean communities shared with us. I would not be able to jump right into one of those local ensembles, nor would I have conceived the same engineering solutions employed by some of those communities. I also realized that I could not use my own "tools" to play some of the traditional music we heard. A local Bolivian musician sat down with us to share a song on his *moseño*. After a couple of minutes, Bernhard and I attempted to join him, but we struggled to match his sound, both the tone and the pitch. This is because our horns were built from and for a completely different musical context.

This can happen in engineering as well, as many Western engineers bring tools such as remote-sensing devices and satellite positioning systems to remote areas and cannot get accurate readings due to lack of signal, rendering them unable to produce a mapping of the site that compares with local documentation of topography. With the training we receive, it can be easy to observe technology that is different from our own and assume it is insufficient or inferior. Alex reflects on this very scenario in her essay when discussing the dome structure fabricated from earth and straw in Janko Kollo. Our automatic reaction was to think that it was a haphazard construction out of necessity when in fact, it achieved many of the standards we seek in Western countries, such as durability and ease of assembly, at far less cost. Having seen this assumption disproved in both music and engineering during our travels strengthens the argument that Western ideas and technology are not always superior or appropriate for every environment.

Power Dynamics: Letting Go of Our Ego

The perception that Indigenous knowledge and practices are inferior is largely fueled by the deeply rooted power dynamics between Western and non-Western countries. This power dynamic manifests as a pressure and tendency to view ourselves as the "expert" in these interactions and demonstrate our knowledge or experience. In engineering, this can often result in ineffective solutions because we worry that we are hurting our reputation by pursuing a design that seems less advanced or less "developed", which can make us hesitant to embrace techniques or technologies from non-industrialized communities. In music, this can result in the inability to meaningfully make music together or share traditions. I certainly experienced this in Bolivia when I had to confront my bias of Indigenous music and my lack of comfort with new experiences. Another reason I was afraid to improvise or interact with local musicians on my horn is that I feared messing up or not sounding "good". I also noticed this when I took the class on improvisation—I felt a need to prove my status and worth on horn because of my background in classical music. My experience had taught me that there are "right" and "wrong" notes, and I would get stuck trying to figure out the "right" thing to contribute during collective improvisation. However, when I improvised on an instrument with which I had little to no experience,

I was suddenly much more comfortable joining in and adapting on the spot because I wasn't worried about my reputation.

In both cases, true competence is evident when a professional has the dexterity to pull from relevant knowledge or experience based on the information they are actively taking in. Through Contextual Engineering, I have learned that the expertise of an engineer will shine if they have the skill to reframe the problem from "how can I fix this for them?" to "how can my expertise or resources strengthen their approach to this problem?" For example, in Sierra Leone, we toured the Moyamba District campus of Njala University, specifically meeting with people from some of the projects being conducted in the agricultural and biological engineering department. A doctoral student showed us various test plots where different crops were being grown and explained that the university was working on algorithms to better predict the weather for farming (Fig. 10.4). This illustrates an understanding that weather patterns are changing and becoming less predictable, which causes lower crop yields. While the term "climate change" may not have a strong presence in their rhetoric, they clearly recognize it is happening and are working to address it. An effective engineer would seek this kind of information and figure out where their own expertise fits. However, oftentimes it can be easier to see that they do not have the same understanding of climate change that we may have and assume that we need to assert our own. We instead need to allow space for their knowledge to inform a solution rather than looking for a knowledge deficit that hinders the solution we already had in mind.

Fig. 10.4 Crop test plots for research at Njala University in the Moyamba District, Sierra Leone

In Sierra Leone I also had the opportunity to experience this concept in music. A couple of days, we improvised with Kene Muadie while he played his accordion and sang. I noticed that while Bernhard and I might have a vast knowledge of repertoire on horn, we had the foresight to use only the most relevant knowledge instead of trying to play the best or most impressive piece for our collaborator. This mental preparation allowed us to musically converse more meaningfully and productively with him. This happened when playing with the National Dance Troupe as well. We first listened, and then we thought how we could fit within their sound using what we know.

Experiences in Academia

The catalyst of this project was the way that these two disciplines have presented themselves in academia. I think it speaks to my educational upbringing that my biases about Indigenous knowledge were so implicit that I didn't recognize them. I remember taking human geography in high school, where I learned about "more-developed countries" and "less-developed countries" and did not think to question the assumptions behind these classifications. I wanted a career somewhere in the non-profit sector specifically so that I could help "catch up" the rest of the world to our state of development. This assumes that other countries should be, and desire to be, like us. In music, this same bias was at play whether I knew it or not, as I viewed Western classical music as the most advanced, most impressive thing I could be doing in music. While other kinds of ensembles are interesting, they should take a back seat to the top bands and orchestras, I thought.

Education was provided a bit differently in the countries that we visited than the process to which we were accustomed. In Bolivia, musicians would simply hand us their instruments and assume we would figure it out on our own. This differed from my experience in music, where there is a rigid process for learning an instrument. Even when I would ask my Bolivian instructors through an interpreter whether I was holding or playing the instrument correctly, they usually would not react. One of the community members explained that youth in the community learn how to make music by watching their elders and jumping in as they get comfortable. There are not any formal lessons or teacher/student relationships. I also noticed this learning pattern outside of music, when we heard testimonies from community members that they used a certain technology or technique because they watched the generation before them do it. This seems to be present in Sierra Leone as well, as Philip mentions in his essay how different trades and music traditions are kept alive through the passing down of knowledge through generations, which plays an important role in preserving cultural heritage. Kene also supports this point earlier in the book when he discusses how Mende music is a medium for him to express the traditions and values of his upbringing and convey his stories to others.

When Contextual Engineering brought to my attention the assumptions I was making about the rest of the world, I sought out everything I could to enhance my understanding of how technical design can adapt to different conditions. However, it felt a little bit like swimming upstream. The Contextual Engineering course, the most influential engineering course of my undergraduate experience, was not eligible to count towards my degree plan on the basis that it was not technical enough. When choosing technical electives, it was difficult to find any that added the dimension of societal considerations and applications. I get the sense this challenge is felt for music students as well, although I do not have degree requirements for music. I have heard colleagues mention that they are required to participate in a traditional classical music ensemble and a chamber ensemble every semester. By the time a music student has met all their ensemble requirements, little room is left to experience other kinds of music, or improvisation, without taking it upon themselves to add to an already full schedule.

Challenging the Status Quo

It was Contextual Engineering that taught me to be open-minded towards other methods for approaching design, particularly Indigenous methods that may have certain cultural meaning behind them or be the most effective and resourceful in a given area. My definition of "innovation" has broadened beyond the image of the newest, most cutting-edge technology. Similarly, it was the Improvisers Exchange Ensemble that taught me to be more open-minded about other musical genres and to see them as having equal merit to Western classical music. I used to view the large classical music ensembles as the most important thing I could do in music, but I have grown to realize that those ensembles are just one of the many important musical endeavors I can explore in my education. It does not mean that the classical ensembles are less valuable, it just means that other forms of music can hold that same level of value.

The common ground that shifted my mindset in both of these disciplines is this idea of improvisation—a Contextual Engineering approach and an improvised music approach both open the gates of uncertainty for what can happen in the progression of a technical or musical collaboration. In order to embrace contextual design, I had to get more comfortable with going "off script," and the idea that I may not have all of the answers in the beginning. However, this contradicts one of the metrics in determining an effective engineering design: the more that uncertainty can be reduced, the more attractive an option will look, and there are multiple methods for quantifying levels of uncertainty. In music, I had to get more comfortable with starting a piece without knowing exactly what I was going to play or what would be asked of me. This contradicts the long-established process in Western classical music whereby performers are repeatedly drilled towards a predetermined standardized sound. I also learned to venture away from the common techniques on horn and towards unconventional ways of making sound, such as putting aluminum foil over the bell to get a more metallic, buzzing kind of tone. Even in composition I have become more adventurous and recently used recordings of ambient equipment sounds in one of my engineering fieldwork classes as additional instrumentation for one of my musical experiments.

Arriving at this point required me to overcome the standards or protocols that I had been taught through most of my education. The persistent drive for perfection in my musical upbringing conditioned me to fear failure and "wrong" notes. My education in engineering had taught me that there is a certain design process that needs to be followed, and technical information takes precedence over other information. The way I approach my work in both fields has changed rather dramatically since I started challenging the mindset I was taught in my education. This year in the Improvisers Exchange Ensemble, I led an improvisation with the class in an attempt to musically represent humanitarian engineering work wherein an outside entity visits a community with the objective of building infrastructure. With half of the ensemble representing the "community" and the other half representing the "organization", this exercise provided a unique space for me to reflect on stakeholder interactions in engineering projects. My decision to remove the bounds of knowledge exchange between engineering and music is already proving itself to be beneficial towards my academic journey in each.

This investigation has strengthened my understanding of how to conduct myself as an engineer and a musician in a way that is welcoming to multiple perspectives that are critical to understand. Ignorance of these perspectives can hinder progress in the global conversation surrounding sustainability goals, as the wellbeing of "less-developed countries" is placed in the hands of Westernized society at the expense of constructive input from valuable Indigenous knowledge and unique cultural identities. While I could have studied this issue separately within each discipline, it expedited my understanding of this multifaceted topic to utilize my knowledge within both. My willingness to attempt music improvisation was motivated by the profound effect I had experienced in engineering from being willing to improvise. Witnessing how critical it was to have listening and adaptive skills during collective music improvisation was an endorsement that engineers could benefit from those same skills during collective engineering design.

I consider myself very privileged to have had this opportunity to challenge the status quo I have come to know in both engineering and music. Just seeing the parallels between these two vastly different fields is strong evidence towards the benefit of cross-disciplinary collaborations. Pulling from the experience of different backgrounds or expertise can facilitate greater understanding of a problem, much like how this exploration has led to a deeper understanding of the interplay between Western elitism and Indigenous knowledge and practices. In an increasingly complex world, global challenges will be better addressed if we look beyond our own expertise to see what we have in common with others and how that can be leveraged to solve problems.

Afterword

The Consilience Project has been a journey of discovery, self-reflection, some pain, and a great deal of joyful exploration of new art and technology experiences for our entire team. We say *pain,* because this project has forced us to confront our own Western habitus—our collective understanding of the world built upon implicit biases that we can't even identify most of the time. But we challenged that habitus at every turn in our travels. We recall the debates held in a guest house of the Njala University main campus in Sierra Leone, where we thoughtfully pushed back on each other's perceptions and responses to the stimuli we encountered. We acknowledge the uneasiness that accompanied our realization that the scattered bones among which we were walking in Culli Culli Alto had once belonged to the ancestors of Aymara, who lived in the region more than a thousand years ago, an unimaginable condition given our own death rituals but one that did not generate so much as a blink of the eye among the deeply spiritual Andean indigenous community.

We could not have undertaken this endeavor without the colleagues and guides who escorted us through their homelands, helping us to open our own minds at the same time as they were afforded a new perspective on their own contexts. The conversations that occurred throughout our experiences together kept us all delightfully off balance, and for this we are grateful, because it is from that tangle with uncertainty that we were able to examine our own conditions of comfort, why they exist, and how they prevent us from exploring new ways of thinking, designing, and performing.

Not surprisingly, this investigation has generated far more questions for us than it resolved. What do we do with this new information and insight? Can we possibly apply what we have learned on our travels to our lives, our research, and our teaching? And, at the end of the day, should we?

Our shared and individual experiences in academia would have us believe that The Consilience Project was an amusing mental exercise but one that has no applicability to *real* scholarship in our respective disciplines, a reflection of the narrowness of perspective that brought us together in the first place. If anything, our explorations have strengthened

A.-P. Witmer et al. (eds.), *Consilience*, Synthesis Lectures on Engineers, Technology, & Society, https://doi.org/10.1007/978-3-031-58399-5

our collective resolve to shake Western intellects out of a smug inertia by demonstrating that there are place-based knowledges, technologies, and art forms that we scholars from the privileged caste ignore simply to reinforce our own positions as self-sufficient arbiters of all things sophisticated.

Ching and Creed (1997) frame this beautifully in the introduction to their book, *Knowing Your Place: Rural Identity and Cultural Hierarchy*:

> Essential secrets of power often lurk in the last place where you would think to look. Finding them is inevitably difficult, but the value of seeking lies in the possibilities for self-determination that these secrets promise. We thus propose looking in the places that are culturally the most remote...

We have begun to look in these places, and we have found that our assumed position of superiority is delusional and does damage to ourselves as well as to the world's societies whose culture, identities, and values differ from our own. While none of us seeks power from this exploration, it is our hope that the power to respect, engage with, and learn from non-industrialized societies will emerge with the realization that the characterization of *Developed Countries* versus *Developing Countries* is a product of our own making.

Through consilience, we have learned that *development* is not a pre-ordained path that others must follow to achieve our standard of excellence and authority. Truly developed societies recognize that every population evolves in its own way, following its own path, while employing tremendous capabilities and talents that are critically valuable to the context in which they were derived.

Reference

Ching, B. & Creed, G. W. (1997) *Knowing your place: rural identity and cultural hierarchy.* New York: Routledge.